IEEE
Std 446-1974

IEEE Recommended Practice for Emergency and Standby Power Systems

Sponsor

Industrial and Commercial Power Systems Committee
of the
Industry Applications Society

Library of Congress Catalog Number 74-13049

Foreword

In 1968 the Industrial and Commercial Power Systems (I&CPS) Committee within the Industry and General Applications Group (IGA) of the Institute of Electrical and Electronics Engineers (IEEE) recognized that a need existed for a publication which would provide guidance to users and suppliers of emergency and standby power systems.

The nature of electric power failures, interruptions, and their duration cover the range in time from microseconds to days. Voltage excursions occur within the range from 20 times normal or more to a complete absence of voltage. Frequency excursions vary as widely in the form of harmonics to direct current.

These variables occur due to a multitude of conditions, both in the power system ahead of the point of the user's service entrance, and following the service entrance within the user's area of distribution.

Such causes as lightning, automobiles striking power poles, ice storms, tornadoes, switching to alternate lines, and equipment failures are but a few of the causes of variables in the electric power supply ahead of the service entrance.

Within the user's area of distribution are such items as shorted and open circuits, undersized feeders, equipment failures, operator errors, temporary overloads, single phasing, unbalanced feeders, fire, switching, and many other generators of variables.

In the past the demand for reliable electric power was less critical. If power was completely interrupted too often, another source was found. If voltage varied enough to cause a problem, a regulator or a larger conductor was installed.

As processes, controls, and instrumentation became more sophisticated and interlocked, the demand developed to shorten the length of outages. Increased safety standards for people required emergency and exit lighting. Many factories added medical facilities which needed reliable electric power.

With the advent of solid-state electronics and computers the need for continuous, reliable, high-quality electric power became critical. Many installations required uninterruptible power, virtually free of frequency excursions and voltage dips, surges, and transients.

In 1969 a working group was established under the Industrial Plants Power Systems Subcommittee of the I&CPS to collect data and produce a publication entitled "Emergency Power Systems for Industrial Plants." Later that year the scope of the work was enlarged to include standby power since in meeting various needs the two systems were often found to be intertwined or one system served multiple purposes.

As the work progressed it became apparent that industrial and commercial needs contained more similarities than differences. Systems available to supply the required power for industry were found applicable to both fields. Once again the scope of the work was expanded by including commercial requirements. Finally the working group was changed to the status of a subcommittee under the I&CPS, and the proposed publication was directed toward establishing recommended practices.

Subcommittee Members and Contributors

Donald E. Woods, *Chairman*
Donald McWilliams, *Secretary*

Graydon M. Bauer	Wilfred Henschel
Don Bigbie	Walter R. Kettenring
Rene Castenschiold	Donald T. Michael
Kao Chen	Pat O'Donnell
G. E. Comeau	Donald W. Zipse
Joseph T. Danko	T. O. Zittel

Ingeborg M. Stochmal, *Editorial Consultant*

Contents

1. Scope

This publication presents the recommended engineering principles and practices for the selection and application of emergency and standby power systems. Industrial and commercial user needs and recommended solutions to those needs are detailed.

The effects of electrical disturbances are a threefold problem, requiring close cooperation between the industrial or commercial user, the electric utility, and the equipment manufacturer. This work has been written from the user's viewpoint and directed toward his need. But this is not possible to present without also detailing the vital parts played by the other two members. Each is a leg of a triangle necessary to complete the whole figure.

Those responsible for reliable plant operation will find general operational needs itemized with the recommended technical parameters of the electric power supply required.

Commercial facility designers, operators, and owners will find operational needs listed, with references to mandatory requirements of laws, regulations, codes, and standards.

Electrical utility companies can utilize the "General Need Guidelines" in identifying the continually changing user requirements for continuous power within close tolerances. From this knowledge further assistance to the user can be provided to meet his needs.

Equipment manufacturers and system engineers will find information in the "General Need" section which will assist them in designing the necessary hardware and systems to fulfill the requirements.

Specific technical and economic information is included on available equipment and systems with recommendations to meet the requirements for various types of installations.

Within the content, the user will find the answers to the following questions:

(1) Is an emergency or standby power system or both needed, and what will it accomplish?

(2) What types of systems are available and which can best meet the needs?

(3) What is the estimated purchased and installed price range of the various systems per kilovolt-ampere?

(4) What are the operating and maintenance requirements for maintaining system reliability?

(5) Where can additional information be obtained?

Although some needs and solutions overlap, the following fields have been excluded: aircraft, government, marine, military, mining, and utility.

2. Definitions

Definitions followed by ♦ were taken from IEEE Std 100-1972, Dictionary of Electrical and Electronics Terms (ANSI C42.100-1972). All other terms have been defined specially for use in this standard.

availability. The fraction of time that a system is actually capable of performing its mission. ♦

commercial power. Power furnished by an electric power utility company; when available, it is usually the prime power source.

computer. (1) A machine for carrying out calculations. (2) By extension, a machine for carrying out specified transformations on information. ♦

controller, automatic (process control). A device that operates automatically to regulate a controlled variable in response to a command and a feedback signal. ♦

data processing. Pertaining to any operation or combination of operations on data. ♦

data processor. Any device capable of performing operations on data, for example, a desk calculator, a tape recorder, an analog computer, a digital computer. ♦

dropout voltage (or current). The voltage (or current) at which a magnetically operated device will release to its de-energized position.

emergency power system. An independent reserve source of electric energy which, upon failure or outage of the normal source, automatically provides reliable electric power within a specified time to critical devices and equipment whose failure to operate satisfactorily would jeopardize the health and safety of personnel or result in damage to property.

firm power. Power intended to be always available even under emergency conditions. ♦

forced outage. A power outage that results from failure of a system component requiring that it be taken out of service immediately, either automatically or by manual switching operations, or an outage caused by improper operation of equipment due to human error.

frequency droop. The absolute change in frequency between steady state no load and steady state full load.

frequency regulation. The percentage change in frequency from steady state no load to steady state full load, which is a function of the engine and governing system:

$$\%R = \frac{F_{nl} - F_{fl}}{F_{nl}} \times 100$$

harmonic content. A measure of the presence of harmonics in a voltage or current waveform expressed as a percentage of the amplitude of the fundamental frequency at each harmonic frequency. The total harmonic content is expressed as the square root of the sum of the squares of each of the harmonic amplitudes (expressed as a percentage of the fundamental).

load shedding. Intentional, selective disconnection of specific loads from their source of power, for the purpose of reducing the load on a system.

power failure. Any variation in electric power supply which causes unacceptable performance of the user's equipment.

power outage. Complete absence of power at the point of use.

prime power. That source of supply of electric energy utilized by the user which is normally available continuously day and night, usually supplied by an electrical utility company but sometimes by the owner generation.

redundancy. Duplication of elements in a system or installation for the purpose of enhancing the reliability or continuity of operation of the system or installation.

scheduled outage. A loss of electric power that results when a component is deliberately taken out of service at a selected time, usually for purposes of construction, preventive maintenance, or repair.

standby power system. An independent reserve source of electric energy which, upon failure or outage of the normal source, provides electric power of acceptable quality and quantity so that the user's facilities may continue in satisfactory operation.

transient. The change in a variable during transition from one steady-state operating condition to another.

uninterruptible power supply (UPS). A system designed to provide power, without delay or transients, during any period when the normal power supply is incapable of performing acceptably.

utility power (see commercial power).

3. General Need Guidelines

While all who use electric power desire perfect frequency, voltage stability, and reliability at all times, this cannot be realized in practice. Within a complex facility, the requirements are continually changing and becoming more demanding and interlocked.

Power interruptions are caused by (1) natural conditions such as storms, floods, and earthquakes, (2) man-made conditions such as automobiles crashing into power poles and vandalism, and (3) material and equipment failures both in plant and out of plant.

Power failures occur due to voltage and frequency excursions outside of established standards for which the equipment was designed. These variations occur due to short and open circuits, motor starting, switching, static electricity, load growth beyond the designed capacity of the system, and numerous other causes.

Lightning, wind, and rain produced by thunderstorms cause power failures in the form of power interruptions and transients. Fig 1 is useful in determining the probability of such power failures depending upon the user's geographic location. However, most utilities or commercial power companies through the use of proper shielding and surge arrestors can provide service that is of equal reliability in any area.

It is recommended that measurements be made of the voltage and frequency excursions on the existing power system prior to finalizing the need, and in order to determine the type of emergency and standby power system to be installed to fill this need. At this writing equipment capable of recording the magnitude and duration of transients down to the microsecond range is rapidly becoming available for purchase or rental at a reasonable cost. These instruments vary from digital peak reading memory voltmeters to more sophisticated chart recorders of the digital and analog type and cathode-ray display units. 1972 purchase prices are in the range of $900 to $28 000.

Systems installed should be selected to overcome the measured troublesome transients as well as longer interruptions, all of which fall under the general category of "power failures."

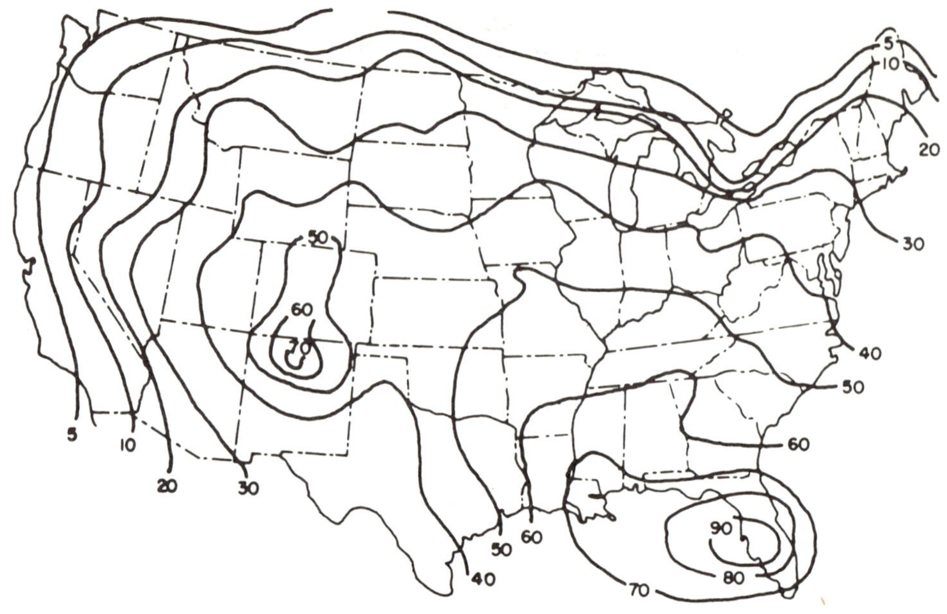

**Fig 1
Average Number of Thunderstorm Days per Year**

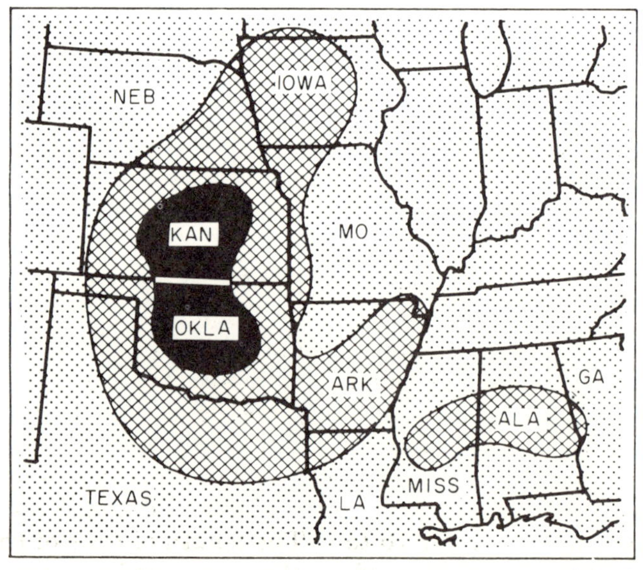

**Fig 2
Approximate Density of Tornadoes. Cross-hatched
areas — 100–200 in 40 years; black area –
over 200 in 40 years [1]**

Table 1
State Codes and Regulations for Emergency Power in the United States

	Alabama	Alaska	Arizona	Arkansas	California	Colorado	Connecticut	District of Columbia	Delaware	Florida	Georgia	Hawaii	Idaho
DOES THIS STATE HAVE LEGISLATION FOR EMERGENCY POWER	Yes	No	No	Yes	Yes	Yes	Yes	Yes	No	No	Yes	Yes	Yes
TYPES OF BUILDINGS AND/OR INSTALLATION SITES COVERED **	A			F	A B	A B	A B C D F G	J			A B C D E F	A D	A
FOR (A) LIGHTS ONLY (B) LIGHT & POWER	A			B	B	B	A	A			B		B
POWER SOURCE REQUIRED ***	D			B	B C A	D	A	C			A B	A B	B
APPLICABLE GOVERNING AGENCY *	7			5	7 5 8	7	5	2			9	5	7

	Illinois	Indiana	Iowa	Kansas	Kentucky	Louisiana	Maine	Maryland	Massachusetts	Michigan	Minnesota	Mississippi	Missouri
DOES THIS STATE HAVE LEGISLATION FOR EMERGENCY POWER	Yes	No	No	No	No	Yes	Yes	Yes	Yes	Yes	Yes	No	No
TYPES OF BUILDINGS AND/OR INSTALLATION SITES COVERED **	A B					A B	J	A K	A B D F	A B	A		
FOR (A) LIGHTS ONLY (B) LIGHT & POWER	B					B	A	A	A	B	B		
POWER SOURCE REQUIRED ***	A B C					B	A B C	A C	D	B	B		
APPLICABLE GOVERNING AGENCY *	7					5 7	5	6 5	4 6 7 8	5	7		

	Montana	Nebraska	Nevada	New Hampshire	New Jersey	New Mexico	New York	North Carolina	North Dakota	Ohio	Oklahoma	Oregon	Pennsylvania
DOES THIS STATE HAVE LEGISLATION FOR EMERGENCY POWER	Yes	Yes	No	Yes	Yes	Yes	Yes	No	Yes	No	No	Yes	Yes
TYPES OF BUILDINGS AND/OR INSTALLATION SITES COVERED **	A	A B D G	K	A B C D	A B C D F	A D	A B		A B			A B	J
FOR (A) LIGHTS ONLY (B) LIGHT & POWER	B	B	B	A	B	B	B		B			B	B
POWER SOURCE REQUIRED ***	A B	B C	A B C	A	A B	A B	A B		B			A B	A B C
APPLICABLE GOVERNING AGENCY *	7 5	7	3	5	9	7	7		7 5			7	8

	Rhode Island	South Carolina	South Dakota	Tennessee	Texas	Utah	Vermont	Virginia	Washington	West Virginia	Wisconsin	Wyoming
DOES THIS STATE HAVE LEGISLATION FOR EMERGENCY POWER	Yes	Yes	Yes	Yes	Yes	No	No	Yes	Yes	Yes	Yes	No
TYPES OF BUILDINGS AND/OR INSTALLATION SITES COVERED **	D	A B	A B	J	A H C			A C D E F	A B D	A B	J	
FOR (A) LIGHTS ONLY (B) LIGHT & POWER	A	B	A	B	B			A	A	B	B	
POWER SOURCE REQUIRED ***	D	D	B	A B C	B			D	A B	A	D	
APPLICABLE GOVERNING AGENCY *	5	7	7	8	4 7			5	5 8	7	8	

* 1 = Attorney General
2 = Dept. of Licenses & Inspections
3 = Advisory Board
4 = Department of Public Safety
5 = Fire Marshall
6 = Department of Public Buidlings
7 = State Board of Health
8 = Department of Labor & Industries
9 = Not Specified

** A = Hospitals
B = Nursing Homes
C = Schools
D = Theaters & Public Gathering Places
E = Office Buildings
F = Hotels & Apartment Buildings
G = Airports
H = Fire & Police Stations
J = All Public Buildings, State & Comm'l
K = State Buildings only

*** A = Battery Powered Lights
B = Emergency Generator Sets
C = Alternate Source of Electricity from another Feeder
D = None specified

From Electrical Generating Systems Marketing Association, Report on State Codes and/or Regulations for Emergency Power (Feb 16, 1965; updated Jul 22, 1971).

NOTE: Nevada has no legislation covering emergency power. However, emergency power installations are a policy requirement of the State Planning Board for all state-financed public buildings and are encouraged in all other public facilities.

Fig 2 shows the density of tornadoes in several states. These violent local storms take their toll of power systems, as do more widely located storms such as hurricanes, ice and snow storms, and floods. The weighted consideration must be based on their frequency and severity in the user's location.

For much electrical equipment 70 to 90 percent voltage dip protection is adequate, that is, transfer to emergency power when the line voltage drops to 70 percent of normal and retransfer when it returns to 90 percent. When specialized loads are used, the manufacturer's recommendation should be followed. For motor loads, the protection can be increased to provide 80 to 90 percent protection. Too close a voltage spread will cause many false or unnecessary transfers; too wide a spread will cause equipment damage and malfunctions. The length of time of the voltage dip is of paramount importance.

Justification of the expenditures necessary to fulfill "user needs" for emergency and standby power systems falls into three broad classifications:

(1) Mandatory installations to meet laws and regulations of federal, state, county, and city

(2) Maintenance of safety and health of people involved during a power failure

(3) Increased profits due to fewer and shorter power failures

Power for the safety of personnel and to prevent pollution of the environment may become increasingly mandatory. A continual check of laws and regulations must be maintained to be current with the requirements. Public Law Number 91-596, Williams-Steiger Occupational Safety and Health Act of 1970 (OSHA), may materially alter the electric power reliability and availability requirements as the full interpretation and application of its provisions are published. Refer to Section 6 for listings of sources of additional information.

Table 1 is a guide to state codes and regulations for emergency power systems in the United States. All latest codes and regulations for the area in which the industrial or commercial facility is located must be consulted and followed.

Installations must conform to the requirements in NFPA No 70, National Electrical Code (1975) (ANSI C1-1975) as related to emergency and standby systems (see Articles 517, 700, and 750). Essential electrical systems for hospitals are also required [NFPA No 76A, Essential Electrical Systems for Health Care Facilities (1973)].

Table 2 lists the needs in thirteen general categories with some breakdown under each to indicate major requirements. Ranges under the columns "Maximum Tolerance Duration of Power Failure" and "Recommended Minimum Auxiliary Supply Time" are assigned based upon the experience of the subcommittee members. Written standards have been referenced where applicable.

In some cases under the column "Type of Auxiliary Power System" both emergency and standby have been indicated as required. An emergency supply of limited time capacity may be used at a low cost for immediate or uninterruptible power until a standby supply can be brought on line. An example would be the case where battery lighting units come on until a standby generator can be started and switched on to the critical loads.

Following the table, the "General Need" listings are presented in greater detail with recommendations as to the type of equipment or system which should be used.

Users of this publication may wish to skip the detailed presentation of each "General Need" and go directly to Section 4, "Systems and Hardware." If so, care should be taken that all individual needs have been recognized and listed so that suitable power systems can be selected to meet all requirements.

Readers using this recommended practice may find that various combinations of general needs will require an in-depth system and cost analysis which will modify the recommended equipment and systems to best meet the specific requirements.

Small commercial establishments and manufacturing plants will usually find their requirements under two or three of the general need guidelines such as 3.1 Lighting and 3.8 Space Conditioning. Large manufacturers and commercial facilities will find that portions or all of the need guidelines 3.1 through 3.13 apply to their operations and justify or require emergency and backup standby electric power.

Table 2
Condensed General Criteria for Preliminary Consideration

Section	General Need	Specific Need	Maximum Tolerance Duration of Power Failure	Recommended Minimum Auxiliary Supply Time	Type of Auxiliary Power System		System Justification
					Emergency	Standby	
3.1	Lighting	Evaucation of personnel	Up to 10 s, preferably not more than 3 s	2 h	×		Prevention of panic, injury, loss of life Compliance with building codes and local, state, and federal laws Lower insurance rates Prevention of property damage Lessening of losses due to legal suits
		Perimeter and security	10 s	10–12 h during all dark hours	×	×	Lower losses from theft and property damage Lower insurance rates Prevention of injury
		Warning	From 10 s up to 2 or 3 min	To return to prime power source	×		Prevention or reduction of property loss Compliance with building codes and local, state, and federal laws Prevention of injury and loss of life
		Restoration of normal power system	1 s to indefinite depending on available light	Until repairs completed and power restored	×	×	Rick of extended power and ight outage due to a longer repair time
		General lighting	Indefinite; depends on analysis and evaluation	Indefinite; depends on analysis and evaluation		×	Prevention of loss of sales Reduction of production losses Lower risk of theft Lower insurance rates
		Hospitals and medical areas	0.1 s to uninterruptible; Life Safety Code (NFPA No 101) allows 10 s for engine to start and power to be available	To return of prime power	×	×	Uninterruptible service to patients by surgeons, medical doctors, nurses, and aids Compliance with all codes, standards, and laws Prevention of injury or loss of life Lessening of losses due to legal suits

Table 2 (Continued)

Section	General Need	Specific Need	Maximum Tolerance Duration of Power Failure	Recommended Minimum Auxiliary Supply Time	Type of Auxiliary Power System Emergency	Standby	System Justification
		Orderly shut-down time	0.1 s to 1 h	10 min to several hours	×		Prevention of injury or loss of life Prevention of property loss by a more orderly and rapid shutdown of critical systems Lower risk of theft Lower insurance rates
3.2	Startup power	Boilers	3 s	To return of prime power	×	×	Return to production Prevention of property damage due to freezing Provision of required electric power
		Air compressors	1 min	To return of prime power		×	Return to production Provision for instrument control
3.3	Transportation	Elevators	15 s to 1 min	1 h to return of prime power		×	Personnel safety Building evacuation Continuation of normal activity
		Material handling	15 s to 1 min	1 h to return of prime power		×	Completion of production run Orderly shutdown Continuation of normal activity
		Escalators	15 s to no requirement for power	Zero to return of prime power		×	Orderly evacuation Continuation of normal activity
		Conveyors	15 s to 1 min	As analyzed and economically justified		×	Completion of production run Completion of customer order Orderly shutdown Continuation of normal activity
3.4	Mechanical utility systems	Water (cooling and general use)	15 s	½ h to return of prime power		×	Continuation of production Prevention of damage to equipment Supply of fire protection

Table 2 (Continued

Section	General Need	Specific Need	Maximum Tolerance Duration of Power Failure	Recommended Minimum Auxiliary Supply Time	Type of Auxiliary Power System		System Justification
					Emergency	Standby	
		Water (drinking and sanitary)	1 min to no requirement	Indefinite until evaluated		×	Providing of customer service Maintaining personnel performance
		Boiler power	0.1 s	1 h to return of prime power	×	×	Prevention of loss of electric generation and steam Maintaining production Prevention of damage to equipment
		Pumps for water, sanitation, and production fluids	10 s to no requirement	Indefinite until evaluated		× ×	Prevention of flooding Maintaining cooling facilities Providing sanitary needs Continuation of production Maintaining boiler operation
		Fans and blowers for ventilation and heating	0.1 s to return of normal power	Indefinite until evaluated	×	×	Maintaining boiler operation Providing for gas-fired unit venting and purging Maintaining cooling and heating functions for buildings and production
3.5	Heating	Food preparation	5 min	To return of prime power		×	Prevention of loss of sales and profit Prevention of spoilage of in-process preparation
		Process	5 min	Indefinite until evaluated; normally for time for orderly shutdown, or to return of prime power		×	Prevention of in-process product damage Prevention of property damage Continued production Prevention of payment to workers during no production Lower insurance rates

Table 2 (Continued)

Section	General Need	Specific Need	Maximum Tolerance Duration of Power Failure	Recommended Minimum Auxiliary Supply Time	Type of Auxiliary Power System		System Justification
					Emergency	Standby	
3.6	Refrigeration	Special equipment or devices which have critical warmup (cryogenics)	5 min	To return of prime power		×	Prevention of equipment or product damage
		Depositories of critical nature, (blood banks, etc)	5 min	To return of prime power		×	Prevention of loss of material stored
		Depositories of noncritical nature, (meat, produce, etc)	2 h	Indefinite until evaluated		×	Prevention of loss of material stored Lower insurance rates
3.7	Production	Critical process power (sugar factory, steel mills, chemical processes, glass products, etc)	1 min	To return of prime power or until orderly shutdown		×	Prevention of product and equipment damage Continued normal production Reduction of payment to workers on guaranteed wages during nonproductive period Lower insurance rates Prevention of prolonged shutdown due to nonorderly shutdown
		Process control power	Uninterruptible (UPS) to 1 min	To return of prime power	×	×	Prevention of loss of machine and process computer control program Maintaining production Prevention of safety hazards from developing Prevention of out-of-tolerance products

Table 2 (Continued)

Section	General Need	Specific Need	Maximum Tolerance Duration of Power Failure	Recommended Minimum Auxiliary Supply Time	Type of Auxiliary Power System		System Justification
					Emergency	Standby	
3.8	Space conditioning	Temperature (critical application)	10 s	1 min to return of prime power	×	×	Prevention of personnel hazards Prevention of product or property damage Lower insurance rates Continuation of normal activities Prevention of loss of computer function
		Pressure (critical) pos/neg atmosphere	1 min	1 min to return of prime power	×	×	Prevention of personnel hazards Continuation of normal activities Prevention of product or property damage Lower insurance rates Compliance with local, state, and federal codes, standards, and laws
		Humidity (critical)	1 min	To return of prime power		×	Prevention of loss of computer functions Maintenance of normal operations and tests Prevention of explosions or other hazards
		Static charge	10 s or less	To return of prime power	×	×	Prevention of static electric charge and associated hazards Continuation of normal production (printing press operation, painting spray operations)
		Building heating and cooling	30 min	To return of prime power		×	Prevention of loss due to freezing Maintenance of personnel efficiency Continuation of normal activities
		Ventilation (toxic fumes)	15 s	To return of prime power or orderly shutdown	×	×	Reduction of health hazards Compliance with local, state, and federal codes, standards, and laws Reduction of pollution

Table 2 (Continued)

Section	General Need	Specific Need	Maximum Tolerance Duration of Power Failure	Recommended Minimum Auxiliary Supply Time	Type of Auxiliary Power System		System Justification
					Emergency	Standby	
		Ventilation (explosive atmosphere)	10 s	To return of prime power or orderly shutdown	X	X	Reduction of explosion hazard Prevention of property damage Lower insurance rates Compliance with local, state, and federal codes, standards, and laws Lower hazard of fire Reduce hazards to personnel
		Ventilation (building general)	1 min	To return of prime power		X	Maintaining of personnel efficiency Providing make-up air in building
		Ventilation (special equipment)	15 s	To return of prime power or orderly shutdown	X	X	Purging operation to provide safe shutdown or startup Lowering of hazards to personnel and property Meeting requirements of insurance company Compliance with local, state, and federal codes, standards, and laws Continuation of normal operation
		Ventilation (all categories non critical)	1 min	Optional		X	Maintaining comfort Preventing loss of tests
		Air Pollution control	1 min	Indefinite until evaluated; compliance or shutdowns are options	X	X	Continuation of normal operation Compliance with local, state, and federal codes, standards, and laws
3.9	Fire protection	Annunciator alarms	1 s	To return of prime power	X		Compliance with local, state, and federal codes, standards, and laws Lower insurance rates Minimizing life and property damage

Table 2 (Continued)

Section	General Need	Specific Need	Maximum Tolerance Duration of Power Failure	Recommended Minimum Auxiliary Supply Time	Type of Auxiliary Power System Emergency	Standby	System Justification
		Fire pumps	10 s	To return of prime power		×	Compliance with local, state, and federal codes, standards, and laws Lower insurance rates Minimizing life and property damage
		Auxiliary lighting	10 s	5 min to return of prime power	×	×	Servicing of fire pump engine should it fail to start Providing visual guidance for fire-fighting personnel
3.10	Data processing	Program memory	Micro-seconds	To return of prime power or orderly shutdown	×	×	Prevention of program loss Maintaining normal operations for payroll, process control, machine control, warehousing, etc
		Core and disk file storage	Milli-seconds	To return of prime power or orderly shutdown	×	×	Prevention of program loss Maintaining normal operations for payroll, process control, machine control, warehousing, etc
		Humidity and temperature control	1 min	To return of prime power or orderly shutdown		×	Maintenance of conditions to prevent malfunctions in data processing system Prevention of damage to equipment Continuation of normal activity
3.11	Life support and life safety systems (medical field, hospitals, clinics, etc)	X-ray	Milli-seconds to several hours	From no requirement to return of prime power, as evaluated	×	×	Maintenance of exposure quality Availability for emergencies
		Light	Milli-seconds to several hours	To return of prime power	×	×	Compliance with local, state, and federal codes, standards, and laws Preventing interruption to operation and operating needs

Table 2 (Continued)

Section	General Need	Specific Need	Maximum Tolerance Duration of Power Failure	Recommended Minimum Auxiliary Supply Time	Type of Auxiliary Power System		System Justification
					Emergency	Standby	
		Critical to life, machines, and services	Milli-seconds	To return of prime power	×	×	Maintenance of life Prevention of interruption of treatment or surgery Continuation of normal activity Compliance with local, state, and federal codes, standards, and laws
		Refrigeration	5 min	To return of prime power		×	Maintaining blood, plasma, and related stored material at recommended temperature and in prime condition
3.12	Communication systems	Teletypewriter	5 min	To return of prime power		×	Maintenance of customer services Maintenance of production control and warehousing Continuation of normal communication to prevent economic loss
		Inner building telephone	10 s	To return of prime power	×		Continuation of normal activity and control
		Television (closed circuit and commercial)	10 s	To return of prime power		×	Continuation of sales Meeting of contracts Maintenance of security Continuation of production
		Radio systems	10 s	To return of prime power	×	×	Maintenance of security and fire alarms Providing evacuation instructions Continuation of service to customers Prevention of economic loss Directing vehicles normally
		Intercommunication systems	10 s	To return of prime power	×	×	Providing evacuation instructions Directing activities during emergency Providing for continuation of normal activities Maintaining security

Table 2 (Continued)

Section	General Need	Specific Need	Maximum Tolerance Duration of Power Failure	Recommended Minimum Auxiliary Supply Time	Type of Auxiliary Power System		System Justification
					Emergency	Standby	
		Paging systems	10 s	½ h	×	×	Locating of responsible persons concerned with power outage Providing evacuation instructions Prevention of panic
3.13	Signal circuits	Alarms and annunciation	1 to 10 s	To return of prime power	×	×	Prevention of loss from theft, arson, or riot Maintaining security systems Compliance with codes, standards, and laws Lower insurance rates Alarm for critical out-of-tolerance temperature, pressure, water level, and other hazardous or dangerous conditions Prevention of economic loss
		Land-based aircraft, railroad, and ship warning systems	1 s to 1 min	To return of prime power	×	×	Compliance with local, state, and federal codes, standards, and laws Prevention of personnel injury Prevention of property and economic loss

plied by an automatic transfer switch [2]. It is generally considered that an average level of 0.4 fc is adequate where passage is required and no precise operations are expected [3].

Table 3 summarizes the user's needs for emergency and standby electric power for lighting by application and areas.

3.2 Startup Power

3.2.1 *Introduction.* Assume a "cold" boiler and a "dead" plant without electric power or steam. From this premise several very important questions must be answered, such as the following:

(1) How will the plant be protected from freezing in cold weather? Even with gas heaters will there be sufficient heat without fans, and without interlocked make-up air units running?

(2) A steam turbine generator is on hand but without forced draft, induced draft, boiler feed water, flame detectors, or control power. How can it be started?

(3) A gas turbine generator has been installed, but how can this be started without bringing it up in speed with a small steam turbine, an electric motor, or other prime mover? Gas compressors may be necessary, and these also require a prime mover of some type.

(4) Steam and electrically driven fire pumps are out of service. There may be no major fire protection until electric power and steam are restored.

(5) An uninterruptible power supply of sufficient capacity is probably not on hand; otherwise, steam and electric power would not be down.

These statements illustrate the fact that adequate startup power is one of the most important considerations in the original design of any plant. Millions of dollars worth of equipment could be standing idle in a time of critical need if no allowance had been made for starting the machines under unexpected conditions such as a major power outage.

3.2.2 *Example of System Utilizing Startup Power.* Starting major plant equipment without outside power is commonly referred to as a "black start" and is accomplished by using only the facilities available within the plant. One example of a system designed with "black start" capability, with a minimum electrical startup system, would be a large gas turbine driving a centrifugal compressor in the natural gas pipeline industry, where the high-pressure gas from the pipeline is used to drive expansion turbines and gas motors for cranking the turbine, operating pumps, and positioning valves. By utilizing the high-pressure gas for the large horsepower requirements, a small engine-driven generator, fueled by natural gas, is used to provide electric power for turbine accessories, battery charging, providing lights, and powering other critical loads. When the turbine is running, a shaft-driven generator provides larger quantities of electric power for all station requirements; and the small generator is placed on "standby."

3.2.3 *Lighting.* In the design of the startup power system, first consideration should be given to installing battery-operated lights in the vicinity of the standby power source and switchgear.

3.2.4 *Engine-Driven Generators.* The battery-cranking power for the engine may also be used for some lights. Cranking may also be accomplished by compressed air supplied to a tank by the plant compressed-air system and prevented from leaving the tank into the plant system by a check valve.

The engines may be sized for short-term operation only if a gas, oil, or steam-turbine-driven generator or other type of continuous power generation can be brought onto the line after startup. If no generation as a prime source of electric power has been installed, the diesel or engine generator should be sized for supporting all electrical needs for the generator auxiliaries, boilers, critical emergency lights, fire signals, exit lights, and other items listed in Table 2.

3.2.5 *Battery Systems.* Special consideration should be given to the design of the plant battery system or uninterruptible power supply, allowing (1) adequate battery capacity to provide power for the necessary startup control systems, following a programmed safe stop; and (2) special disconnecting devices to automatically disconnect large power-consuming systems from the battery system, when possible, to conserve battery capacity for restart.

With capacity for these minimum facilities in operation, consideration may then be given to installing sufficient capacity at the same time to support additional justified needs.

3.2.6 *Other Systems.* Mobile equipment may suffice if it can reasonably be assumed to be available when needed. (Who has the highest priority when all have the need?) An alternate standby public utility line may also be available at a low cost from a separate source of supply. A neighboring plant with live generation may assist in an emergency.

3.2.7 *System Justification.* A definite workable plan with the proper equipment should be evaluated, and action taken as justified, prior to the need.

3.3 Transportation

3.3.1 *Introduction.* This topic covers the moving of people and products by methods which depend upon electric power; the importance of maintaining power ranges from desirable to critical.

3.3.2 *Elevators.* Where two or more elevators are in use in buildings three or more stories high, the elevators or banks of elevators should be connected to separate sources of power. There are situations where standby power is required for all elevators within 15 s. Savings may be made by supplying power during outages of the normal supply to one half the elevators installed, providing the traffic can be rerouted and the capacity of the elevators is adequate. Power must be transferred to the second bank of elevators within 1 min or so of the prime power loss to clear stalled elevators. Power may be left on this bank until normal power returns.

Where elevator service is critical for personnel and patients, it is desirable to have automatic power transfer with manual supervision. Operators and maintenance men may not be available in time if the power failure occurs on a weekend or at night.

(1) *Typical Elevator System.* Fig 3 shows an elevator emergency power transfer system whereby one preferred elevator is fed from a vital load bus through an emergency riser while the rest of the elevators are fed from the normal service. By providing an automatic transfer switch for each elevator and a remote selector station, it is possible to manually select individual elevators, thus permitting complete evacuation in the event of power failure. The engine generator set and emergency riser need only be sized for one elevator, thus minimizing the installation cost. The controls for the remote selector, automatic transfer switches, and engine starting are independent of the elevator controls, thereby simplifying installation.

(2) *Regenerated Power.* Regenerated power is a concern for motor generator type elevator applications. In some elevator applications, the motor is used as a brake when the elevator is descending and generates electricity. Electric power is then pumped back into the power source. If the source is

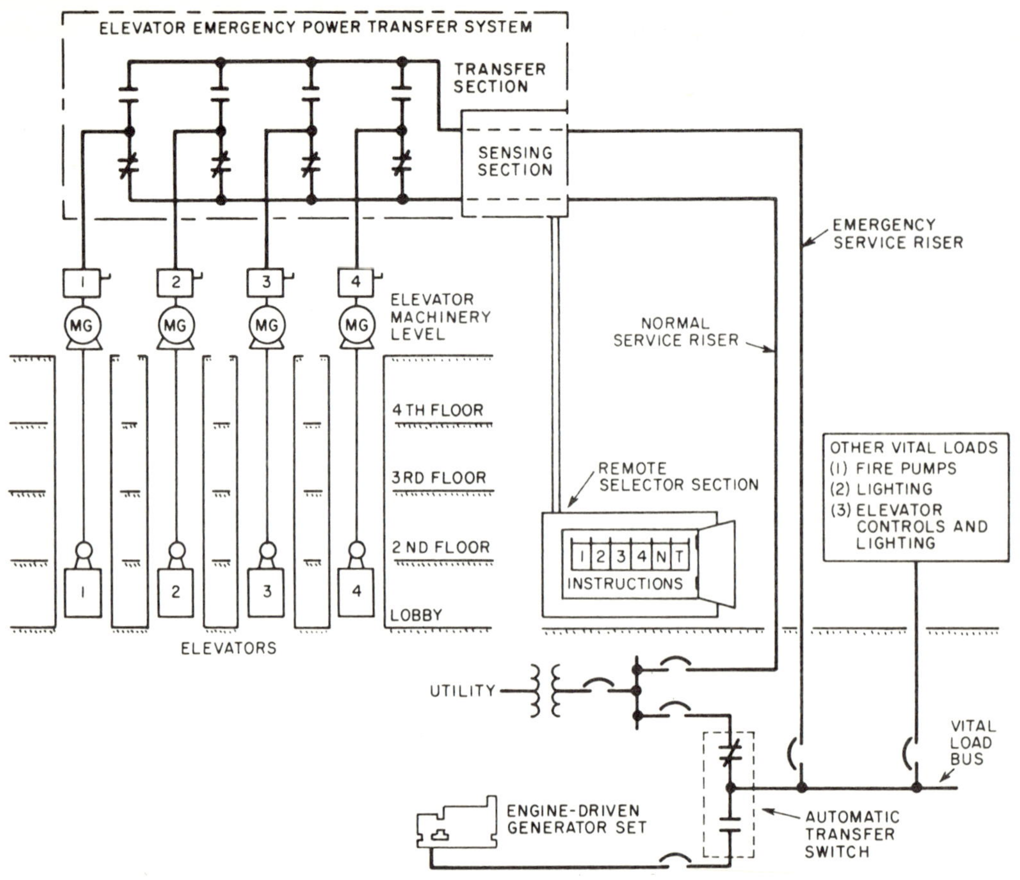

Fig 3
Elevator Emergency Power Transfer System

commercial utility power, it can easily be absorbed. If the power source is an engine-driven generator, the re-generated power can cause the generating set and the elevator to overspeed. To prevent overspeeding of the elevator, the maximum amount of power that can be pumped back into the generating set must be known. The permissible amount of absorption is 20 percent of the generating set's rating in kilowatts. If the amount pumped back is greater than 20 percent, other loads must be connected to the generating set, such as emergency lights or "dummy" load resistances. Emergency lighting should be permanently connected to the generating set for maximum safety. A dummy load can also be automatically switched on the line whenever the elevator is operating from an engine-driven generator.

3.3.3 Conveyors and Escalators. Escalators and personnel conveyors may require emergency power since physically handicapped persons ride up and would have great difficulty walking down, even though a normal person would be able to do so were the power off.

Those who have used conveyors or elevators in a small way for a time and who have continued to grow, should be wary and check their needs for standby power. To feed a few cattle or gather a few eggs using conveyors may have once eased someone's labor, but labor could always fill in if needed. As operations become larger and larger, there is a point where an outage of electric power could produce a disaster. The gathering of 250 000 eggs a day, or the feeding of 10 000 head of cattle is dependent upon power when needed.

3.3.4 *Other Transportation Systems.* Power for charging equipment for battery-powered vehicles is usually a non-critical requirement. Time available for the emergency generation to come on line varies from several minutes to several hours.

A few examples of transportation systems which may need standby power are listed below:

(1) Conveyors for raw materials through to finished goods

(2) Warehouse high-stacking loading and unloading equipment and conveyors delivering finished goods to shipping facilities

(3) Slurry pumps for long pipe lines

(4) Livestock feeder conveyors

3.4 Mechanical Utility Systems

3.4.1 *Introduction.* Often the need for mechanical utilities is as great as that for electric power. These are interdependent. We can speak of the utility systems as a whole since, to most managers and corporate officials, these are all united in a group under the heading "Utilities."

3.4.2 *Typical Utility Systems for which Reliable Power May Be Necessary.* Mechanical utility systems comprise the following services for which reliable electric power is usually required:

(1) Compressed air for pneumatic power

(2) Cooling water (including return pumps, pressure pumps, tower fans)

(3) Well water or other pumped source for personnel use

(4) Hydraulic systems (200 lb, 1500 lb, or other pressures as required)

(5) Sewer systems (sanitary, industrial, storm)

(6) Gas systems (natural, propane, oxygen), including compressors

(7) Fire pumps and associated water supplies

(8) Steam systems (low and high pressures)

(9) Ventilation (building and process)

(10) Vacuum systems

(11) Compressed air for instrumentation

Additional systems may exist in some plants, but the list will alert a plant manager or engineer to the various needs and potential losses should electric power not be available.

Systems may be required for manufacturing and services to maintain other services. For example, electric power, water, and compressed air for boilers used to supply steam for the generation of electricity.

A 0.1 to 5 s power failure may cause operating engineers to spend minutes or hours restarting equipment and making adjustments until all systems are again stable. Such disruptions to production should be prevented if economically justifiable.

3.4.3 *Orderly Shutdown of Mechanical Utility Systems.* An orderly shutdown may be acceptable and can be provided with a short-time smaller supply source of power for the following requirements:

(1) To maintain temperature or pressure on vulcanizers until the product can be finished.

(2) To maintain hydraulic pressure until a batch process is completed, or

until the pressure can be released without loss.

(3) To operate sump pumps for a time until all process water has been shut off or has drained back into the sumps to prevent flooding. This becomes serious when the water is contaminated with oil and other waste when flooding occurs and special handling may be needed to comply with antipollution requirements.

(4) To maintain ventilation to clear explosive atmospheres while a normal shutdown proceeds. Purging air is critical to some oven-drying processes.

(5) To prevent sanitary sewers from overflowing before personnel evacuation takes place.

(6) To run gas compressors for the finishing of a critical process.

A complete listing has not been intended, but representative needs of various types should be enough to alert a plant engineer so that a plan may be prepared for an orderly shutdown. Management may then act on the needs justified without being caught unprepared in an emergency situation. Although management may not approve the recommendations immediately, the plan will be ready for resubmitting, approval, and implementation when an interruption occurs, alerting management to the need for an emergency or standby power supply.

3.4.4 *Alternates to Orderly Shutdown.* An orderly shutdown may not be acceptable. Then an alternate may be selected as follows:

(1) Maintain utilities without an interruption. Full capacity standby power must be available as well as uninterruptible power for boiler controls, on-line computers, and essential relays and motor starters.

(2) Accept an outage but return on standby power. Full capacity power must be available. Startup of functions will be required since magnetic motor starters and relays will have dropped out. Although power is off for 0.1 to 5 s, it will take 15 min to ½ h or more for return to normal operation.

3.5 Heating

3.5.1 *Maintaining Steam Production.* Continuous-process plants require uninterrupted steam production. Minimum requirements for continuous steam production are sufficient combustion air, air to instruments and actuators, water and fuel supplies, plus a continuous power supply to most flame supervision systems. The maximum interruption tolerable is that duration during which the inertia of the fans or pumping equipment will maintain flows and pressures above minimum limits. Table 4 illustrates how this can be achieved.

3.5.2 *Process Heating.* Process heating is defined as heat required to maintain certain process materials at the required temperature. Noncritical heating processes, due to the inherent heat capacities of such systems, can withstand a power interruption of considerable duration; say, 5 min to a maximum of several hours.

Any of the following systems would be adequate for this application:

(1) Engine-driven generators, off line

(2) Multiple utility services on line and relayed, or off line, switches, and transfers

(3) Turbine (combustion, steam, water) off line or on line

(4) Mechanical stored-energy system, with auxiliary motor generator sets off line

Other heating processes such as cord and fabric treating and drying are of such a critical nature that loss of heat will cause an out-of-specification product within 10 s, but the gas or oil burners and flame detectors are sensitive to

Table 4
Systems for Continued Steam Production

Components	Allowable Outage Duration	Systems
Flame supervision systems	Nil	Mechanical stored energy systems Motor generator set ride-through UPS systems
Motor controls and instrumentation	Nil	Same as above
Boiler fans	½ to 2 s	Multiple utility services either on line and relayed, or off line, switched, and transferred
Air compressor	To 30 min, depending on storage; nonessential air users should be automatically shut off	Multiple utility services Turbine- or engine-driven generator, off line Turbine (combustion, steam, water) off line
Water pump	To 5 min, depending on water drum capacity and upsets in steam production caused by power disturbance	Multiple utility services either on line and relayed, or off line, switched, and transferred Turbine (combustion, steam, water) off line, automatic start
Oil pumps (for burners)	To 15 cycles, more with flywheel	Multiple utility services, on line and relayed Turbine (combustion, steam, water) on line
Electric oil supply pumps	Several minutes	Turbine- or engine-driven generator, off line Multiple utility services, on or off line

drops in voltage of about 40 percent for a second or less. In this case, an uninterruptible supply for the controls and an engine generator for a 10 s main power supply or an alternate feed system may be required.

Infrared drying of enamel on automobiles and appliances is a form of heating by electric energy which must be maintained. A short interruption, perhaps 10 s, may be acceptable to bring on-line standby generation or perform automatic switching to an auxiliary power source.

Losses may be substantial should power be lost during the heat treating of metals and when either direct or indirect melting of metals is in process. Two evaluations must be made, one based on not losing the flame on the fuels used, and the other based on accepting an interruption for a short period with a restart necessary. Uninterruptible power is several times more expensive than switching or emergency power, but saves the product and prevents process interruption. Switching or emergency power is less expensive but product or process losses may exceed the initial savings in the chosen electric supply system.

Induction and dielectric heating are other forms of electrical heating which may or may not allow short interruptions to be tolerated. Most processes, whether in production or in other fields, could tolerate an interruption of sufficient duration to come on line with switched sources or an engine- or turbine-driven generator.

3.5.3 *Building Heating.* Buildings, even in the coldest regions, can usually be without heating for a minimum of 30 min. The same systems listed for noncritical heating processes will be suitable for this application.

3.6 Refrigeration

3.6.1 *Requirements of Selected Refrigeration Applications.* Requirements for refrigeration are usually noncritical for short power interruptions of several minutes to several hours. The need may become extremely critical as the length of time of the outage increases. Consider these refrigeration needs:

(1) Production of ice cream or the freezing of foods may stop in the middle of the process. Not only will all production be lost during a power failure, but damage may result to the product in process.

(2) Material in storage may be in jeopardy as temperatures rise. Cafeterias, frozen food lockers, meat cooling and storage facilities, dairies, and other food operations require refrigeration and will soon be in trouble as the length of the power outage increases.

(3) Scientific tests of long duration may require accurately maintained low temperatures. Short outages of electric power may destroy the tests and require repeating. An expensive and time-consuming process should be provided with a standby power system.

(4) Medical facilities require refrigeration for blood banks, antibiotics, and for long-range laboratory experiments and cultures which could be spoiled.

(5) As the state of the art of superconductivity develops, improves, and spreads to applicable fields, power for cryogenic refrigeration equipment operation will probably become critical.

Where present refrigeration units are electrically driven, when new units of moderate size are to be installed in permanent locations, and when other needs exist (as is usually the case) for emergency or standby power, a common engine-driven generator or alternate utility source should be considered.

Because of the slow rise in temperature of cold-storage facilities, a savings may be made by the use of a smaller than normal standby electric generator. By switching power to various units in turn, an acceptable storage temperature may be maintained until normal electric power has been re-established.

3.6.2 *Refrigeration to Reduce Hazards.* Certain chemical processes are exothermal and release heat during the chemical reaction. Loss of the cooling or refrigeration system may cause severe damage or even an explosion.

3.6.3 *Typical System to Maintain Refrigeration.* A manual starting of an engine-driven generator, turbine, or alternate utility supply will usually suffice, providing a suitable alarm is installed to notify responsible persons of a loss of refrigeration.

3.7 Production

3.7.1 *Justification for Maintaining Production in an Industrial Facility.* Loss prevention in production facilities due to a power failure is justified on the total sum of many tangible and intangible savings. Some of these items to consider follow.

Is there a guaranteed wage clause in the labor contract? If so, there will be a direct loss in wages paid for which no production is received. Sometimes a small electric supply system can be justified supplying finishing, inspection, office, and other areas where most of the people work, but where power requirements are low, and leaving the heavy-power demanding machines shut down until normal power is returned.

Who is waiting for the product? There are periods when the products are being routed to warehouses, and machines are not running at capacity. In this case the cost of machine down time is lower than it would be for a product which a customer will not receive or will receive late if production is at full capacity and an outage occurs.

What is the cost of product spoiled in process? In the rubber industry material may become sealed in vulcanizers, extruders, or mixers at high temperatures and will be difficult and expensive to remove. Steam or water pressure may drop to zero and prevent proper cures with losses due to poor or ruined products.

If all electric power is lost during certain processes in the making of sugar, glass, steel, pharmaceuticals, rubber, paper, chemicals, and some other materials, the product must be scrapped.

What is the cost of consequential damages? While the material ruined may be scrap, there may be as many problems and costs associated with its removal as with the loss itself. Some material must be dug out, or removed by hand, piece by piece, until lines are cleared or chambers are empty and clean so that an orderly startup can follow.

What is lost in reestablishing work efficiency? A ½ h electric power interruption disorganizes the workers. Experience indicates that, following the interruption, it will take men and women at least two or more hours to settle down, go to work, and reach the production level at which they were operating just prior to the power failure. It may take days to reestablish normal procedures in scheduling of incoming and outgoing materials and in telegrams, letters, notices, and calls explaining happenings and changing promises.

A less tangible item lost is good will. For example, in the film processing industry the customer may not consider the replacement of the exposed film with unexposed as adequate compensation for his "once in a lifetime" pictures which were spoiled due to a power failure.

Real and potential costs and losses must be calculated or estimated and added together to justify an emergency and standby power system for industrial and commercial facilities.

A reasonable estimate of the costs associated with each past power failure should be calculated and recorded in a journal with the date, duration, and conditions existing at the time. As time goes on this will be found to be valuable factual backup information for budget requests.

3.7.2 *Equations for Determining Cost of Power Interruptions.* A rough estimate of the cost of down time in an industrial plant from a cash flow viewpoint may be gained by substituting the proper figures in the following example:

$$E = AD(1.5B + C)$$
$$= 5 \times 3(1.5 \times \$4 + \$3) = \qquad \$135$$

$$H = FG$$
$$= 250 \text{ yd} \times \$1.30 = \qquad 325$$

$$I = JK(B + C) + LG$$
$$= 2 \times 6(\$4 + \$3)$$
$$+ 400 \text{ yd} \times \$1.30 = \qquad 604$$

Total cost of the 3 h
power failure $\qquad \$1064$

where

A = Number of productive employees affected (assumed 5)

B = Base hourly rate of employees affected, in dollars (assumed \$4/h)

C = Fringe and overhead hourly cost per employee affected, in dollars (assumed \$3/hr)

D = Duration of power interruption, in hours (assumed 3 h)

E = Cost of direct labor, in dollars

F = Units of scrap material due to power failure (assumed 250 yd)

G = Cost per unit of scrap material, in dollars (assumed \$1.30/yd)

H = Scrap loss due to power failure, in dollars

I = Cost of startup, in dollars

J = Startup time, in hours (assumed 2 h)

K = Number of employees involved in startup (assumed 6)

L = Units of scrap material due to startup (assumed 400 yd)

After the cost of down time has been calculated, the savings in utilities should be subtracted to arrive at a total cost of down time.

3.7.3 *Commercial Buildings.* For commercial establishments a similar example may be assembled based on the length of the power interruption, labor cost, loss of profit on sales, loss due to theft, and startup costs.

3.7.4 *Additional Losses Due to Power Interruptions.* In addition to losses relating to cash flow are those more difficult to calculate but which should be included when available and applicable such as

(1) Prorated depreciation of capital costs

(2) Depreciation in quality in process materials

(3) "Cost" of money invested in unused materials or machines

Other losses may occur under special or unusual conditions. In an industrial plant operating at 100 percent capacity any loss in production results in the loss of the profit of the item or service. The prorated cost of fixed and variable overhead becomes a loss. Customers may switch to competitors. Expenditures for standby power have additional justification under this condition.

3.7.5 *Determining Likelihood of Power Failures.* Next must be determined the likelihood of a power failure by studying the record of the plant or utility company electrical supply, or by transmitting the service requirements to the local utility and obtaining their recommendations. Examples of recorded power failures are shown in Table 5.

Rather than complete power failures as recorded in Table 5, Table 6 covers short-term dips.

A projection should be made, working with the utility company, as to whether the power reliability will improve or decline. Since the cost of a power failure, as defined in this publication, is paid by the user, it is important that he relate the reliability of power duration and quality to the need and justification.

3.7.6 *Factors that Increase Likelihood of Power Failures.* As full designed load is reached or exceeded, the probability of a power failure increases. A similar probability exists as systems become more complex and as system equipment becomes older.

3.7.7 *Power Reserves in the United States.* Power reserves in the user's area should be investigated. Reports published in January, 1972, show the following winter peak power reserves by regional areas in the United States:

Northeast	29.5 percent
Southeast	18.6 percent
East Central	21.1 percent
South Central	64.8 percent
West Central	27.8 percent
West	19.2 percent

Table 5
Example of Recorded Power Failures

Date	Time	Duration	Transmission Line
9 March	09:52	10 min	14
11 June	21:53	12 s	14
11 June	22:13	9 s	14
15 July	20:40	5.5 s	13 +22
17 July	19:13	1 – 2 min	14 (9 times)
	20:44		

Table 6
Example of Recorded Short-Term Dips

Date	Time	Line	Duration (cycles)	Line-to-Ground Dip (per unit) E_a	E_b	E_c	Voltage After
14 April	21:50	32	18	0.86	0.81	0.75	1.0
30 April	15:53	System	43	0.83	0.92	0.92	1.0
9 May	07:52	23,24	32	0.90	0.85	0.83	1.0
9 May	07:54	20	24	0.31	0.31	0.58	1.0
9 May	07:55	22	46	1.00	1.00	0.69	1.0
						0.40	
						0.50	
						0.06	
17 May	00:43	13, 22	21	0.50	0.69	0.31	1.0

Sixteen areas within these regions are below 10 percent. These figures were filed with the Federal Power Commission by the nation's electric utilities [4]. Adequate reserve margins above anticipated peak-load demands provide a guide to power reliability as the margin provides for some contingencies.

3.7.8 *Examples of Standby Power Applications for Production.* The following typical examples were taken at random of users who found, after evaluation, purchase, and installation, that a standby power system was justified.

(1) A drug company installed a special engine generator backup system based on the need for constant temperatures in manufacturing processes. The value of processes saved during blackouts soon exceeded the cost of the equipment installed by more than 10 times.

(2) A photographic film-processing company specified a 45 kW engine generator when the film-processing machines were ordered. Film must be moved along in each aspect of the process within 30 s or loss of quality will result.

(3) A 400 kW, 480 V, three-phase engine generator was installed in a highly automated egg farm. Electric service continuity was mandatory for egg production and the welfare of the "machines" (chickens). During the first year two electric service outages occurred and the automatic emergency generator reliably supplied power.

3.7.9 *Types of Systems to Consider.* An engine- or turbine-driven generator or an alternate independent utility source usually will fulfill the requirements for standby power. More than a standby system may be required for critical loads. Motor starters, contactors, and relays held closed by a coil and magnetic structure are especially sensitive to short-time power outages or voltage dips. Their drop-out characteristics vary with respect to voltage level and the length of duration of the voltage dip. As a guide, a voltage dip to 70 or 60 percent of rated voltage for 0.5 s will drop out many of the devices. The longer the time of the voltage dip and the greater the voltage dip, the more devices will drop open. Depending on the application of these devices, an emergency or uninterruptible power supply system may be justifiable, especially where boiler controls, critical chemical processes, safety devices, and other critical systems are required to be maintained [5].

Devices much more likely to fail or misoperate due to voltage excursions are modern static switching devices which have almost no time delay, and pressure, temperature, and flow transmitters and receivers of the electronic type. Automatic calculators for process control, data reduction, and logging are in this same category [6].

An uninterruptible power supply system for critical on-line computer or boiler control loads may be required since a 30 cycle outage may need restarting the plant. In combination the uninterruptible power supply system may be of short duration since the standby equipment should be designed to be on line in from 10 to 60 s.

Factory clocks used for production control, or for the basis of incentive wage payments, should be arranged to maintain accurate time by the use of batteries for power during the time of any power interruption.

Support of production facilities will justify most or all of the user's needs detailed in Section 3. The evaluation, jus-

tification, and decision to purchase and install a standby, emergency, or uninterruptible power supply system, or a combination of these systems, must include the consideration of all the electric power requirements for all listed needs in case of a power failure.

3.8 Space Conditioning

3.8.1 *Definition.* Space conditioning is a controlled environment, either to maintain standard ambient conditions or some artificial alteration of a standard environment in a building, room, or other enclosure.

3.8.2 *Description.* A controlled environment may include any of the following variables:

(1) Temperature
(2) Vapor content
(3) Ventilation
(4) Lighting
(5) Sound
(6) Odor
(7) Gas
(8) Dust
(9) Organisms

Specific conditions in commerce and industry require control of any number of these variables to maintain the required conditioned environment. Most installations depend upon electric power for maintaining proper conditions. Refer to Sections 3.1 and 3.5 for additional information on lighting and heating.

3.8.3 *Examples of Space Conditioning where Auxiliary Power May Be Justified.* Typical situations and facilities in which a comprehensive study of the need for supplemental electric power is warranted are the following.

(1) Commercial and laboratory horticultural botanical installations may require programmed cyclic control of temperature, humidity, and lights to develop the crop yield or desired experimental results. The loss of temperature or

humidity control for 6 to 8 h can result in a total loss of a crop. A greenhouse must have maintained temperature to produce.

(2) Tropical animal raising requires control of ventilation, temperature, humidity, and lighting. All are completely dependent on electric power. A loss of heat or cooling can result in death or illness to all animals being raised. Lighting and temperature changes from the established cycle can induce unwanted breeding periods in many exotic creatures. Egg production may be greatly curtailed by temperature changes or loss of light.

(3) Agricultural operations are often located in remote areas where the long utility lines are susceptible to damage.

(4) Final operations and packaging of materials susceptible to contamination are conducted in "clean room" type environments. A power interruption will shut down the total operation, and contamination may result as people exit, unless the room is kept under positive pressure to prevent in-drafts from bringing in contaminants. Such contaminants will necessitate a complete recleaning of the room before it can be used again.

(5) In large process plants priorities may be established for essential loads. Nonessential loads may be dropped automatically by load-shedding systems in the event of a power failure, and other limited emergency sources must be relied upon. A reevaluation of these priorities should be undertaken. Critical temperature controls may have been placed in an air-conditioned space. These controls must function to achieve an orderly process shutdown, but can fail due to overheating caused by the lack of ventilation or cooling.

(6) Windowless buildings or inside rooms may become unsafe for occupancy during an extended power interruption. Power supplies to keep these areas ventilated should be installed if evacuation of all personnel is not acceptable.

3.8.4 *Typical Auxiliary Power Systems.* Needs may require any one or more types of emergency and standby power systems available. An engine-driven generator or combustion turbine should be considered for these applications. Switching to a separate alternate electric utility line would involve a lower capital cost, should such a line be readily available. In addition, a short-time uninterruptible power supply system may be required for critical applications to supply power during engine startup or switching.

3.9 Fire Protection

3.9.1 *Codes, Rules, and Regulations.* Various codes, standards, laws, rules, and regulations contain either advisory or mandatory statements related to emergency and standby power systems for fire protection. See Title 24, California Administrative Code, Part 3, Basic Electrical Regulations (article E700, Emergency Systems); NFPA No 101, Life Safety Code (1973); IEEE Committee Report [7] (pp 19, 20); and Katz [8]. There appears to be more written concerning the wiring system reliability and needs than on the source of electric power supply or on the checking and maintenance of the complete installation. All three are vital.

The requirements of all local, state, and national standards and codes should be determined. The insurance company who will underwrite the insurance coverage can provide valuable assistance in making sure that all requirements are met.

A common-sense approach should be used even beyond meeting the letter of the law and insurance requirements as a minimum standard. The real goal is to avoid a destructive fire, or in the event a fire does start, that it be held to a local area with minimum damage to property and no harm to personnel. Knowledge in depth of the industrial facility and processes by plant engineers and other responsible plant personnel should be utilized to reduce the likelihood of a fire and the extent of damage should one begin.

3.9.2 *Arson.* Arson may be a source of a fire and the need of power for plant security, lighting, signaling, and communication covered in other sections should be regarded as contributing to the reduction of possible loss by fire from this cause.

3.9.3 *Typical Needs.* The user's possible specific needs connected with the general need of emergency and standby power systems for fire protection include the following:

(1) Power, usually batteries, to crank the engine driving the fire pump

(2) Sprinkler flow alarm systems

(3) Communication power to notify the fire department and to assist in guiding their activities

(4) Lights for the firemen to work by in the buildings, around the outside area, and mobile on company trucks

(5) Power for the boilers which supply steam-driven fire pumps

(6) Motors driving fire pumps, well pumps, and booster pumps

(7) Air compressors associated with fire water tanks

(8) Smoke and heat alarms

(9) Deluge valves

(10) Electrically operated plant gates, drawbridges, etc

(11) Communications such as public-address systems for directing evacuation of personnel

3.9.4 *Feeder Routing to Fire Protection Equipment.* Electric power distribution systems supplying fire equipment should be routed so as not to be burned out by a fire in the area they are protecting.

3.10 Data Processing

3.10.1 *Definition.* By definition (IEEE Std 100-1972), data processing pertains to any operation or combination of operations on data.

3.10.2 *Description.* In general, data are gathered in analog or digital form and converted to one or the other system. These data in electrical form are then processed through one or several formulas by an electronic computer. The results are traced or printed out, or a signal is generated and fed back to form a closed-loop system which provides control to match preset conditions. Commonly both types of outputs are available and used.

3.10.3 *Justification for Maintaining Power to Data-Processing Equipment.* Both industrial and commercial users apply data-processing systems with a computer, on call, or on line, using "real time." All systems malfunction should the electric power supply digress from normal by a sufficient amount for a sufficient time. Thus a reliable alternate power supply is necessary to minimize down time.

Both industrial and commercial systems have critical applications where a failure in the data-processing system can affect the safety of men and equipment. Both may have very extensive financial losses and require hours, days, or weeks to recover from a malfunction.

3.10.4 *Process Control and Other Applications.* A short list of control operations which frequently include data-processing (computer) equipment will suffice to alert users of these systems to the hazards and losses that would occur with a power failure which might be a complete interruption or cause a process control malfunction. Some applications fall under both industrial and commercial classifications.

  (1) Industrial applications:
- Materials handling
- Regulating pumping stations at remote locations
- Mixing compounds
- Grinding, drilling, machining
- Twisting textile fibers
- Steel mill processing
- Refining (oil)
- Automatic testing and gauging
- Fabric processing

  (2) Commercial applications:
- Controlling elevators
- Automated checkstand combined with inventory control
- Newspaper production machines
- Airline reservations
- Environmental control
- Analyzing and tabulating of data in banking
- Typesetting
- Accounting
- Traffic control

3.10.5 *History of Developing Needs.* With the advent of the electronic computer as a part of data processing and process control, an increased emphasis was placed on the need for an emergency and standby power system.

Coupled with this need was a superimposed problem which, it was found, could be corrected at the same time, namely, the suppression of most short-time switching interruptions, voltage surges and dips, frequency excursions, and transients. All of these have been a part of the electric power supply in the past, but caused few problems until solid-state electronic equipment came into extensive use.

3.10.6 *Power Requirements for Data-Processing Equipment.* By the proper selection of an electric supply system, the power needs associated with data processing with a computer can be met, namely, a reliable source of electric power at all times and of a higher quality than previously demanded by most devices.

The problem is how to reconcile commercial short-duration power interruptions with shorter and shorter time domains in electronic circuits. A point has now been reached, in present electronic design, where a gap exists between approximately 1.0 μs, the minimum time domain for power interruptions to cause performance effects in electronic equipment, and 5 cycles, the minimum commercial power interruption recovery time. Another gap is also appearing which includes momentary voltage interruptions. For the power industry, discontinuities of less than 15 cycles could hardly be considered a power interruption, yet for many items of equipment, it is a power failure. Such momentary interruptions are rarely reported, so statistics on frequency of occurrence are not readily available.

A way to envision time domains is to note a typical computer effective clock rate of 16 MHz. The clock sequences the electronics. It can be deduced that the computer electronics are sequenced or stepped in 62.5 ns intervals (1 per clock rate of $\frac{1}{16} \times 10^6$ or 62.5×10^{-9} s, or 62.5 ns).

Table 7 shows that the susceptibility in the command system starts at 4 ms (computer memory). Word structure alterations, computer parity errors, memory alteration, and nonprogrammed jumps are among the more serious effects noted as a result of momentary power interruptions. Susceptibility in the display system (first three items in table) started with power interruptions of 1.5 ms. Loss of synchronization of incoming data trains, alteration of memory in decommutators, computers and symbol generators, nonprogrammed jumps, computer parity errors, and

Table 7
Schedule of Time of Short-Duration Power Interruptions Versus Start of Equipment Effects

Line Frequency 60 Hz (cycles)	Time (milliseconds)	Unit	Performance Interrupted	Data Altered	Other Units Affected
Less than 1 cycle	1.5	Decommutator	X	X	
	1.5	Tape recorder			Decommutator
	1.5	Alphanumeric symbol generator	X		Computer
	4.0	Computer external memory	X		Computer
	5.0	Computer typewriter	X	X	Computer
	5.0	Alphanumeric display unit	X		
	5.75	Computer control console	X	X	
	7.0	Card reader	X		Computer
	10.0	Tape search unit	X	X	Computer
	10.0	Card reader	X	X	Computer
	12.0	Computer input–output module	X		Computer
	15.5	Tape controller	X		Computer
Less than 2 cycles	16.5	Compute module	X		
	20.5	Compute module	X	X	
	23.0	External memory	X	X	Computer
	25.0	Receiver decoder		X	Alphanumeric displays
Less than 3 cycles	40.0	Data converter and transmitter	X		Computer
	44.0	Magnetic tape recorder	X		Decommutator
	50.0	Command controllers		X	Control system
Less than 4 cycles	60.0	Scanner		X	
Less than 5 cycles	70.0	Receiver decoder	X		Alphanumeric system
Less than 6 cycles	100.0	Wall clock	X		
Less than 7 cycles	115.0	Magnetic tape recorder	X	X	
	125.0	Magnetic tape recorder	X		Computer
18.5 cycles	200.0	Recorder, chart	X		
18.6 cycles	310.0	Count-down time generator	X		Computers
19.6 cycles	325.0	Count-down time generator	X	X	Computers
24 cycles	400.0	Count-down time generator	X	X	Computers and tape search
27 cycles	450.0	Computer complex console	X		Oscillograph

From Mirowsky [9].

Table 8
Guide to Computer Input Power Quality Parameters by Manufacturer
Early 1971 Data

Company	AC Supply Voltage	Maximum Frequency Variation (Hz)	Maximum Supply Voltage Transient (percent)	Maximum Harmonic Voltage Distortion (percent rms)	Maximum Phase Angle Tolerance (degrees)	Maximum Load Step (kVA)	Maximum Load Unbalance (percent)	Typical Load (kVA)	Maximum Load (kVA)
IBM	208Y/120V or 240V	± 0.5	+ 10, − 8	5	120 ± 6	225	20	240	1500
XDS	208Y/120V or 240V	± 0.5	± 10	15	120 ± 2.5	4	30	25	50
Honeywell	208Y/120V or 240V	± 0.5	± 5	10	120 ± 5	20	15	25	105
RCA	208Y/120V or 240V	± 0.5	± 10	5	120 ± 5	12	3	40	290
Burroughs	220 V ac	± 0.6	−0.7 times nominal for 0.5 s, + 2.5 times for 5 ms	5	Mostly single phase	37.5	Mostly single phase	30	60
Univac	220 V ac	± 0.5	+ 5, − 20 at 10 s intervals	3	Mostly single phase	6	Mostly single phase	13.7	150
Control Data	208Y/120V or 240/480V	± 0.6	± 20 for 30 ms	10	120 ± 5	—	3	100	200

From Allingham [10].

NOTE: Not to be used for design as changes occur frequently.

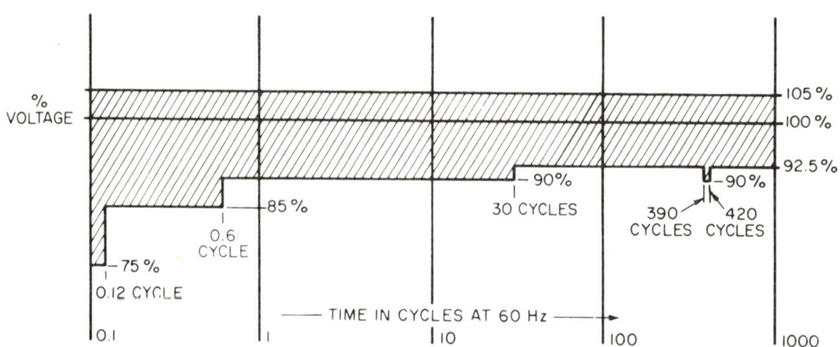

Fig 4
Composite Alternating-Current Line Voltage Tolerances of
Computer Main Frames of Seven Domestic Manufacturers

lockup of computer are among the more serious effects noted as a result of power interruptions.

The results in Table 7 indicate a wider degree of consequences than might actually occur in practice, because coincidence is unlikely between a power disturbance and all performance modes. If it takes 50 μs to transfer data from the computer to a transmission line, it is a question of probability whether a 50 μs event that occurs perhaps 100 times an hour will coincide with a nominal 8 to 10 cycle interruption of approximately 160 ms that might occur twice a day. The chance of coincidence in this example is approximately one in one hundred. There are other functions that occur continuously, such as the storage of data in memory or the decommutating of a serial data train.

Table 8 shows computer input power quality parameters by manufacturer rather than the more detailed data collected from a single installed system in Table 7. The user should consider Table 8 as a guide only since computer designs are in a state of flux and the parameters of power needs are changing rapidly. Fig 4 is a more recent (late 1972) composite of computer manufacturers' volt-

age requirements for reliable performance.

The conventional definition of "interruption" does not include discontinuities and disturbances in the time domains of concern to the electronic system. Therefore, this publication uses the defined term of a "power failure."

3.10.7 *Utility Power.* The supplying utility cannot be expected to provide a perfect power supply because many of the causes of power supply disturbances are beyond the control of the utility. For example, automobiles hit poles, animals climb across insulators, lightning strikes overhead lines, and cyclones, hurricanes, and even high winds blow tree branches and other debris into lines. Although there is less chance for an interruption on an underground system, the duration of any interruption which occurs may be much longer because of the longer time required to find where a cable failure occurred and to make repairs.

Even the operation of protective devices causes power supply disturbances. Overcurrent and short-circuit protective devices require excessive current to operate so there will be a voltage dip on any line supplying the excessive current. Devices opening to clear a lightning flashover and then reclosing cause a momentary outage.

Following is the text of a typical utility sales contract:

"The power delivered hereunder shall be three-phase alternating current at a frequency of approximately sixty (60) hertz and at a nominal voltage of 480Y/277 volts. Except for temporary periods of abnormal operating conditions, variations from normal voltage shall not exceed seven percent up or down. The utility will use reasonable diligence to provide a regular and uninterrupted supply of power, but in case such supply should be interrupted for any cause, the utility shall not be liable for damages resulting therefrom."

3.10.8 *In-House Power Failures.* It is clear that the two systems, the electronic system and the power distribution system, are not compatible in their present configuration or in their specifications (or legal regulations).

Utility companies are by no means the only source of power failures. Failures also occur in the plant system through the loss of power due to short circuits in wiring and failures of local generation including emergency and standby equipment. "Noise" is generated in otherwise acceptable electric power by motors, welders, switches, SCR gating, dielectric heating, short circuits, and by myriad other means.

3.10.9 *How to Maintain Operation of Data-Processing Equipment.* Since an ideal power supply can seldom be obtained from the supplying utility, this text has been prepared to indicate how the effects of power supply disturbances may be reduced to acceptable levels or even eliminated. These possibilities include the following:

(1) Modification of the design of the electronic equipment so as to be impervious to power disturbances and discontinuities.

(2) Modification of the prime power distribution system to be compatible with the electronics.

(3) Modification of both systems to meet a criterion that is realistic for both.

(4) Interposing a continuous electric supply system between the prime source and the computer station and locating it adjacent or near to the computer. This will function as a "buffer" to transients.

This work is concerned with option (4) to meet the user's need, but it is realized that other steps are being taken by computer manufacturers to desensitize their equipment to the transient conditions. This will help in the future, but prime power interruptions will still create a need for emergency and standby power systems.

3.10.10 *Justification of Supplemental Power for Data-Processing Equipment.* Insurance companies provide loss of power and loss of production insurance which can give a cost guideline on how much can be spent to improve the power supply.

Sensitivity of computers to power supply disturbances does not apply to the entire computer system. Power requirements for peripheral equipment such as printers, air-conditioning, lights, and drive motors is much less restrictive than the logic, memory, and control units. For economic reasons the power loads should be separated into those requiring buffer or filter action with a uninterruptible power supply, and those which require no buffering and can accept a 0.5 s to 1 min power interruption while switching to a standby source of electric power.

3.10.11 *Selection of Supplemental Power System.* Because continuously assuring acceptable quality alternating-current power to computer system complexes is so difficult, power buffers are designed to operate between the computer system input terminals and the source of alternating-current power. These power buffers are used whether the power source is a public utility or an in-house generator. In a well-designed system, the generator may be of the standard commercial type and is used not only to power the computer but also the lights and air-conditioning which are essential to computer operation. In a large computer complex the heat output is such that operation without air-conditioning usually cannot be continued much longer than 15 min without damage.

To determine what type of continuous alternating-current power system should be installed, certain criteria must be established. A fundamental consideration is the duration of probable interruptions. A review of future plans for improvement in the alternating-current power distribution system should be made prior to the design and purchase of a system. Table 9 shows computer power buffer performance comparisons by systems.

The rotating mechanical stored-energy system could fulfill requirements of supplying continuous alternating-current power within the tolerances and specifications indicated earlier in this section. This system requires a backup standby generator, or an alternate utility supply source of continuous power with automatic switching to pick up the load before the stored-energy system has dropped below minimum voltage and frequency limits.

Battery systems in conjunction with solid-state input rectifiers and output inverters act as excellent buffers and maintain power during a forced interruption until standby power is available

or, if purchased with sufficient battery capacity, until the prime source of power is restored.

Battery systems with engine generators as standby power in conjunction with a motor generator set act as buffers, provide uninterruptible power of high quality, and may be specified for a short- or long-term supply, depending on the availability of standby power.

3.11 Life Support and Life Safety Systems

3.11.1 *Introduction.* Electric power needs for critical life systems and supporting systems are well documented for hospitals [Title 24, California Administrative Code, Basic Electrical Regulations, and NFPA No 76A (1973)]. Details of the installations required or recommended are covered there and in NFPA No 101 (1973), Katz [8], and NFPA No 70 (1975). Here the concern is from the viewpoint of the user's needs.

3.11.2 *Health-Care Facilities.* Medical and nursing sciences are becoming progressively more dependent upon electrical apparatus for the preservation of life of hospitalized patients. A patient's life may depend upon artificial circulation of the blood. Life is sustained by means of electric impulses that stimulate and regulate heart action. Lighting is needed in strategic areas, and power is needed to safeguard refrigeration.

Interruption of normal electric service to hospitals may be caused by catastrophes such as storms, floods, fires, earthquakes, or explosions when the need for the service is critical. Pyramiding failures on the utility supply system or incidents within the hospital cause loss of power. Electrical systems supplying this equipment must be planned to provide for continuity of vital services at all times.

Table 9
Typical Computer Power Buffer Performance

	Motor Generator and Flywheel	Motor/Flywheel/ Clutch/Generator	Solid-State Uninterruptible Power System	AC Motor/Flywheel Battery/DC Motor/AC Generator
Duration of emergency source	Up to 0.5 s	Up to 15 s	Up to 3 h with batteries; indefinite if engine- or turbine-driven generator is included	For length of battery supply purchased
Voltage regulation	208Y/120V ac ± 1 %	208Y/120V ac ± 1%	208Y/120V ac ± 1%	208 Y/120 V ac ± 2%
Voltage drop or rise for 33 percent load step change from full load	± 8%	± 8%	± 6%	± 10% (50% step)
Voltage transient	0.5 s	0.5 s	0.1 s or less	—
Frequency regulation	60 Hz + 0, −0.5	60 Hz ± 0.5	60 Hz ± 0.1	59.7 Hz ac drive/ 60 Hz ± 0.5 Hz dc drive
Frequency transient	± 0.5 Hz	± 0.5 Hz	None	—
Frequency transient recovery time	0.5 s	0.5 s	No transient	—
Phase angles, unbalanced loads up to 20 percent	120° ± 5°	120° ± 5°	120° ± 1°	—
Harmonic voltage	5% rms maximum	5% rms maximum	5% rms maximum	3% rms maximum
Electromagnetic interference	MIL-1-16910 or better	MIL-1-16910 or better	MIL-1-16910 or better	—

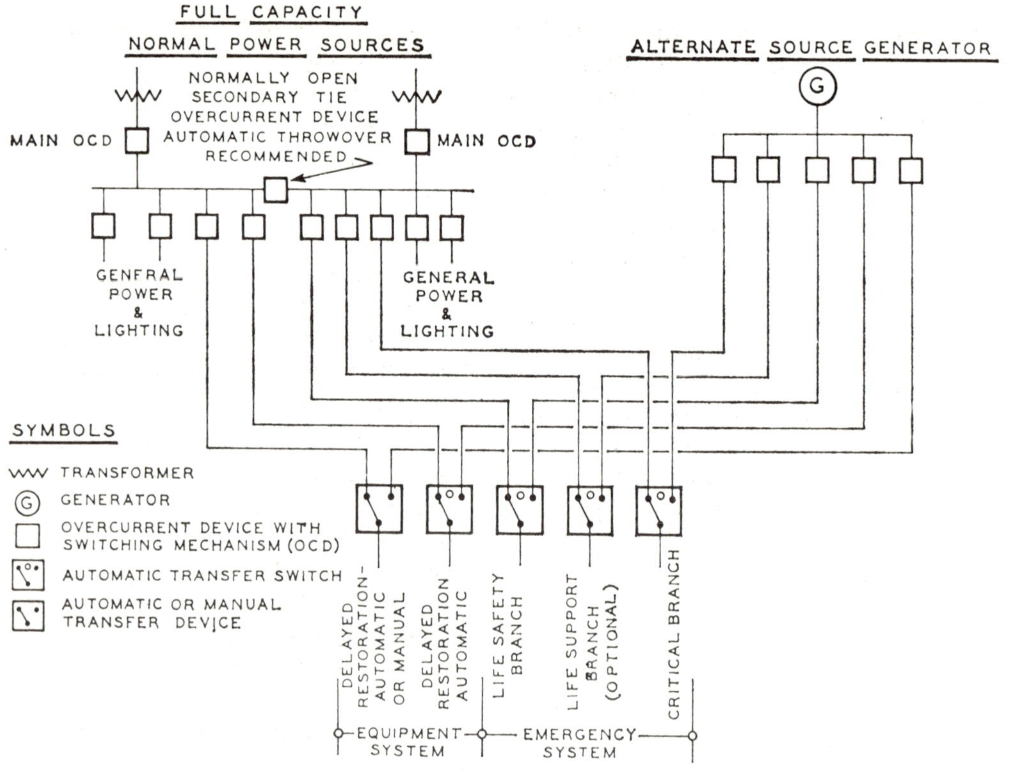

Fig 5
Typical Hospital Wiring Arrangement [NFPA No 70, NEC (1971 ed)]

Since emergency and standby power systems are mandatory in most areas, the state and city building laws and regulations must be checked and insurance and building code inspectors should be consulted.

Periodic tests and maintenance are required to be performed subject to the acceptance of the authority having jurisdiction. A record of these procedures must be kept. Section 5 gives examples of the uselessness of emergency and standby systems which are not maintained and tested to make sure they are operational when the need for them arises.

An example of a typical circuit arrangement for a hospital is shown in Fig

5. NFPA No 76A (1973) includes sources of power, emergency and equipment systems, emergency and critical systems, arrangement of wiring for hospitals and nursing homes, transfer switches, generator sets and prime movers, and maintenance. On switch operation requirements for hospitals it states: "The emergency system and the equipment system shall be so arranged that in the event of failure of the normal power source, an alternate power source shall be automatically connected within 10 seconds to the distribution panels connected to the emergency system and to the time delay and/or non-automatic switching means connected to the equipment system."

Homes for the aged and nursing homes have medical equipment requiring continuous electric power. It is especially recommended that managers of these facilities review their critical requirements. An automatic starting engine generator with automatic switching to critical loads may supply the need economically.

A typical large health-care facility involved in extensive life support type surgical operations may consider a combination arrangement with a static uninterruptible power supply and engine-generator system as shown in Fig 43 as an emergency source of power for certain loads.

It is strongly recommended that the user with a need in this area consult and secure the services of qualified engineers familiar with the special laws, rules, regulations, and common sense needs for power requirements servicing health-care facilities.

3.11.3 *Other Critical Life Systems.* Other critical life systems with similar special laws and regulations applying and which require engineers experienced in providing electric service are as follows:

(1) Controls for pressure vessels such as boilers, and for ovens and similar applications where a failure may lead to an explosion or fire with resulting injury or death

(2) Areas in which there are explosive fumes, dust, powders, and other life hazards [NFPA No 101 (1973)]

(3) Air supply systems for persons in a closed area

(4) Fire pumps, alarms, systems, and telephones

3.12 Communication Systems

3.12.1 *Description.* Communication systems are those facilities which require electric power for verbal, written, or facsimile transmission and reception.

Common systems of this type are
(1) Telephone
(2) Teletypewriter
(3) Paging
(4) Radio
(5) Television

Needs of one or all of the above communication systems during a power failure may well justify the cost of one or more emergency and standby power systems, possibly in conjunction with any other critical loads.

3.12.2 *Commonly Used Auxiliary Power Systems.* Battery or battery and converter equipment are practical sources of power using a float-charging system. Small engine generators are practical and economical. In the size ranges usually required from 1 to 5 kW, the installed cost ranges from about $150 to $300 per kilowatt, depending upon the quality of the equipment, battery life, gasoline storage problems, and amount of automatic equipment deemed to be necessary or convenient.

3.12.3 *Evaluating Need for Auxiliary Power System.* The need of an emergency or standby power system for communications should include satisfactory answers to the following questions:

(1) Will the communication equipment be required

(a) To issue orders for an orderly shutdown of processes and equipment?

(b) To announce instructions to personnel? A typical announcement might be to wait by the machines or to check out for the duration of the shift.

(c) To call for help, issue warnings, and coordinate the work should there be a fire, civil disturbance, vandalism, or other threat to personnel safety or plant security?

(2) How will vital messages be received or sent for remote plants concerning production, inventory, or sales changes?

(3) How will key personnel be found, or instructed; and how will these persons report conditions to a central responsible source of control?

Many other questions may be asked, but the maintaining of communications under emergency conditions will save vital time and expedite the return to normal conditions with reduced confusion.

For industrial plants, the telephone system is usually powered both regularly and on a standby and emergency basis by the telephone company, normally by batteries or by standby generation. In some plants there is a separate in-plant telephone system which may be powered by batteries under a float charge which will maintain the communication system for several hours during a prolonged power interruption. The user should check to see that his system will function for the required time with loss of normal power.

Usually various bells, horns, and other call devices connected to the telephone system are powered by the lighting circuits and will stop functioning if there is a power interruption, even though the receiver and transmitter work. If required, these should be wired to the backup power system.

Teletype equipment functions much the same as the telephone system so that the signal may be present but the printout device will not operate without local power.

Paging systems in plants are frequently extensive, using many hundreds of watts of audio power and several kilowatts of 120 V power. The need to use the system during a power interruption should not be overlooked.

Radio systems are common in industrial plants. While the mobile units for personnel as well as in-plant and out-of-plant car and truck units are generally self-powered by batteries, the main base station usually is connected to the nearest commercial power source. While mobile units may or may not be able to speak to each other, the system as a whole usually stops functioning if the base station power fails. Consideration of emergency and standby power for this base station should not be overlooked.

In some plants and commercial buildings, paging or broadcasting is done by dialing through the telephone system. In this case, both the telephone and paging or radio systems should be supplied with emergency power.

3.13 Signal Circuits

3.13.1 *Description.* A signal circuit supplies power to a device which gives a recognizable signal. Such devices include bells, buzzers, code-calling equipment, lights, horns, sirens, and many other devices.

3.13.2 *Signal Circuits in Health-Care Facilities.* Signal circuits in medical buildings which should be provided with continuous emergency power within 10 s [NFPA No 101 (1973)] include

(1) Fire alarm systems:
    Manually initiated
    Automatic fire detection
    Water flow alarm devices used with sprinkler systems

(2) Alarms required for systems used for piping of nonflammable medical gas

(3) Paging system

(4) Nurse's station signaling system from patient areas

(5) Alarm systems attached to equipment required to operate for the safety of major apparatus

(6) Signaling equipment for elevators in buildings of more than four stories

3.13.3 *Signal Circuits in Industrial and Commercial Buildings.* Signal circuits for commercial buildings and industrial plants which may require continuous emergency power within 1 min include

(1) Fire alarm system

(2) Watchman's tour system

(3) Elevator signal system

(4) Door signals (into restricted areas such as boiler rooms and laboratories with electric door locks)

(5) Liquid level, pressure, and temperature indications

3.13.4 *Types of Auxiliary Power Systems*. The emergency supply for the signal circuits can be (1) engine-driven generators, (2) multiple utility services, or (3) floating battery systems with auxiliary power. Most signal circuits operate down to 70 percent rated voltage and therefore require no special voltage-sensing relays on the transfer device.

It is recommended that an emergency source of electric power be supplied for every part of a fire alarm and security system. A local battery supply on float charge, close to the power need, in continuous service is very reliable. A usually acceptable substitute is an automatic transfer to a battery system when prime power fails.

Signal circuits are usually a small electrical burden and integral part of a total load that also requires an emergency source. Therefore, the selection of emergency system and hardware generally depends upon the requirement of other related loads.

4. Systems and Hardware

4.1 Guidelines for Use. When a user experiences an equipment problem due to failure of the electric power supply, he must either live with the problem, change his equipment or system to perform satisfactorily during a failure of the existing power supply, or alter the supply to prevent potential failures. In many cases, the correct decision is to change the equipment or system, but this publication does not examine those cases.

Once the electric power user's study has shown that the correct approach is to alter or supplement the power supply source, a study should be undertaken to determine the proper systems and hardware which will meet the need at the lowest cost for the electric power required.

This publication describes combinations of systems and hardware which will overcome the following types of electric power failures with good reliability:

(1) Long-time interruption (hours)

(2) Medium-time interruption (minutes)

(3) Short-time interruption (seconds)

(4) Transient interruption (milliseconds)

(5) Over- or undervoltage

(6) Over- or underfrequency

(7) Transients in the prime source of power

(8) Transients caused by the user's equipment

Emergency power systems are of two basic types, (1) an electric power source separate from the prime source of power, operating in parallel, which maintains power to the critical loads should the prime source fail; or (2) an available reliable power source to which critical loads are rapidly switched automatically when the prime source of power fails.

Emergency systems are frequently, but not always, characterized by continuous or rapid availability of electric power of limited-time duration and supplied by a separate wiring system. Frequently the emergency power system has a standby power system available which increases the emergency supply time to as long as needed.

Standby power systems are made up of the following main components:

(1) An alternate reliable source of electric energy separate from the prime power source

(2) Starting and regulating control if on-site standby generation is selected as the source

(3) Controls which transfer loads from the prime or emergency power source to the standby source

It is prudent for the user to establish his practical need from the previous chapters of this publication before specifying and purchasing, since costs will be found to rise as the following demands are specified for systems and hardware:

(1) Longer equipment life

(2) Increased capacity

(3) Closer frequency regulation

(4) Closer voltage regulation

(5) Freedom from voltage or frequency transients

(6) Increased availability

(7) Continuous operation of an uninterruptible power supply system

(8) Increased reliability

(9) Increased temporary overload capability

(10) Quiet operation

(11) Safety from fuel hazards

(12) Pollution-free operation

(13) Freedom from harmonics

(14) Close voltage and frequency regulation with wider range rapid load changes

An allowance should be made for load growth. Future power requirements frequently need to be connected to the emergency and standby system. It may be desirable to add additional existing power loads to the more reliable power bus as soon as the advantages are realized in practice. If additional capacity cannot be justified initially, the equipment and system should be selected and designed for future economic expansion compatible with the initial installation.

Operating costs of the systems and hardware are usually secondary to meeting the need, but should be included as a factor in the selection. These include cost of fuel, inspection frequency, ease of maintenance, frequency of testing, cost of parts, and taxes.

Installation quality should be the best to prevent losing the reliability of electric power designed into the system and purchased in the hardware [11]. It must be guarded against introducing voltage transients into the emergency and standby power distribution system. Satisfactory voltage levels must be maintained under all loading conditions.

For industrial plants, electrical systems should conform to IEEE Std 141-1969, Electric Power Distribution for Industrial Plants. For commercial buildings, electrical systems should conform to IEEE Std 241-1974, Electric Power Systems in Commercial Buildings. Grounding practices should follow the recommendations in IEEE Std 142-1972, Grounding of Industrial and Commercial Power Systems (ANSI C114.1-1973). Additional shielding, bonding, grounding, and even filtering may be required to maintain the quality of the emergency and standby power supply.

When energized sources of power are held available for emergency use, there should be included in the system a light to show that the source is energized and an alarm to signal loss of available power. These should be located where responsible persons can take action, should an alarm sound. An alternate utility source (Section 4.3) or battery system (Section 4.7) are typical on-line sources normally requiring these signals.

Users should take these additional steps to assure performance reliability:

(1) Establish regular inspections using a check sheet and recording exceptions

(2) Perform regular preventive maintenance and repair items of exceptions found during the inspections

(3) Set up a trial at regular intervals simulating a power failure but timed so

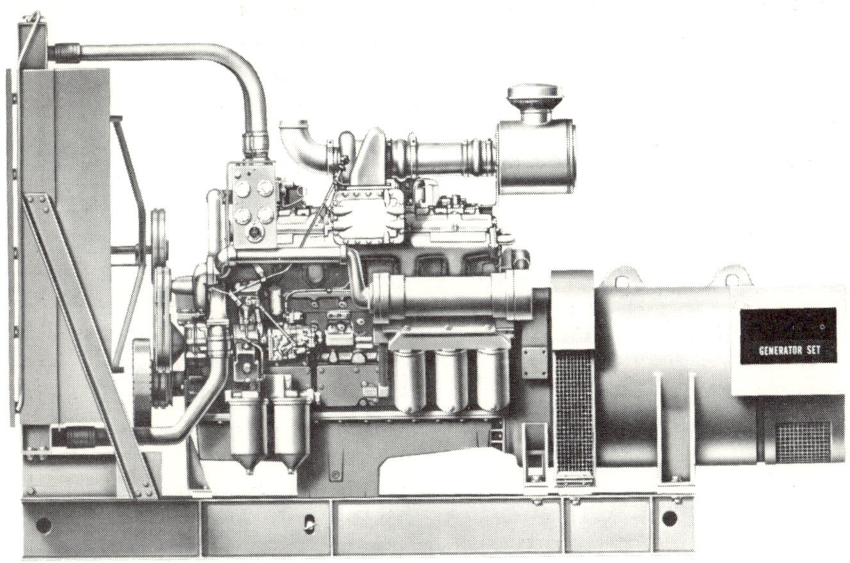

Fig 6
Typical Engine-Driven Generator; Diesel, Gasoline,
or Gas Fueled; 350 W to 2875 kW

as not to encounter hazards or losses should the system not operate as anticipated

There are methods for mathematically determining quantitatively the reliability of an emergency or standby power system. Once the need is established, it may be advisable to calculate the system reliability, especially for emergency systems involving possible injury or loss of life. One such method is detailed in Sawer [12] and Heising and Johnston [13].

In Sections 4.2 through 4.7 the systems and hardware in practical use today are presented. On the horizon are other sources of electric energy and systems which may soon serve to fill some of the needs for emergency and standby power systems. Some of these sources and systems foreseen for future applications are

(1) Fuel cells which convert chemical energy directly into electric energy

(2) Radiant and solar cells which convert radiant energy into electric energy

(3) Chemical luminescence arranged to convert energy into light

(4) Nuclear power generators

(5) Thermocouples which convert heat to electric energy

(6) Radio isotopes which excite chemical panels to produce illumination

Recommended practices for the application of these and other developments will be presented as they become developed to the state of practical application to fill the user's needs as detailed in the previous sections.

4.2 Engine-Driven Generators

4.2.1 *Introduction.* These units are "work horses" which fulfill the need for emergency and standby power. They are available from small 1 kVA units to those of several thousand kilovolt-am-

Table 10
Typical Ratings of Engine-Driven Generators
1971 Prices

Nominal Rating (kW)	Continuous Duty Rating (kW)	Standby Rating (kW)	Power Factor	Prime Mover			Speed (r/min)	Cost (dollars)
				Gasoline	Diesel	Natural Gas/ LP Gas		
5	5	5	1.0	×		×	1800	1040
10	10	12.5		×			1800	1860
25	25	30	0.8		×		1800	4200
100	90	100	0.8		×	×	1800	6000-7000
250	200	250	0.8		×	×	1800	15 000
750	665	730	0.8		×		1200	51 000
1000	875	900	0.8		×		1200	70 000
1000	975(G) 800(D)	1100	0.8		×	×	1200	75 300 67 800

peres. When properly maintained and kept warm, they dependably come on line within 8 to 15 s. In addition to providing emergency power, engine-driven generators are also used for handling peak loads and are sometimes used as the preferred source of power.

They fill the need of back-up power for uninterruptible power systems. Where well regulated systems, free from voltage, frequency, or harmonic disturbances, are required, such as for computer operations, a "buffer" is usually needed between the critical load and the engine-driven generators.

4.2.2 *Diesel-Engine Generators.* A typical diesel-engine-driven generator rated 500 kW is shown in Fig 6. Typical ratings are given in Table 10. Installed operating units with fuel supplies (other than small portable units) will cost between $100 to $170 per kilovolt-ampere. Lower speed units are heavier and more costly, but are more suitable for a continuous supply.

Diesel engines are somewhat more costly and heavier in smaller sizes, but are rugged and dependable. The cost of fuel is lower and the fire and explosion hazard is considerably lower than for gasoline engines. Sizes vary from about 2.5 kW to 4000 kW and above (see Table 10).

4.2.3 *Gasoline-Engine Generators.* Gasoline engines are satisfactory for installations up to about 100 kW output. They start rapidly and are low in initial cost as compared to diesel engines. Disadvantages are a higher operating cost, a great hazard due to the storing and handling of gasoline, and generally a lower mean time between overhaul.

4.2.4 *Gas-Engine Generators.* Natural gas and LP gas engines rank with gasoline engines in cost and are available up to about 600 kW. They provide quick starting after long shutdown periods because of the fresh fuel supply. Engine life is longer with reduced maintenance because of the clean burning of natural gas. However, consideration must be given to the possibility of both the electric utility and the natural gas supply being unavailable concurrently.

To compare directly with gasoline, the natural gas must have a heat value of at least 1100 Btu/ft³. Considerations in selecting natural or LP gas fueled engines are the availability and dependability of the fuel supply, especially in an emergency situation.

4.2.5 *Derating Requirements.* As noted in Section 4.4, altitude will cause a serious derating of the prime mover to deliver the torque required for the full generator output unless a supercharger has been added. Oversizing of the engine is a must for higher altitudes. The generators are less critical in output capacity if adequate cooling air is supplied to carry away the heat generated by the losses.

A general rule for derating engine power loss with altitude increase is to derate about 4 percent for each 1 000 ft increase in altitude. Also, an average derating factor for high ambient temperature is 1 percent for each 10°F above 60°F. Temperature derating is not considered as important as altitude derating.

4.2.6 *Multiple-Engine Generator Set Systems.* Automatic starting of multiple units and automatic synchronizing controls are available and practical for multiple-unit installations. Advantages of several smaller units over one large unit should be considered since emergency and standby power can be available while one unit is being maintained or overhauled. While starting is usually reliable, if the units are warm and maintained by regular exercising, the likelihood of all the units not starting is extremely low as compared to a single unit.

Smaller units also allow the "building block" concept. As capital is made available and the need of increased capacity grows, additional units of identical size and type may be added, thus simplifying the parts, maintenance, and training problems. The trend toward larger emergency power supplies also justifies the use of multiple sets to provide the additional power. Fig 7–10 illustrate typical multiple-set systems.

One very important consideration in the selection of "smaller" versus "larger" units is the service to which they will be subjected. Smaller units are nearly all higher speed (1800 r/min) engine-driven sets; and if these units are to be run continuously for long periods of time, the engine should be evaluated very thoroughly. Experience has shown that many 1800 r/min engine-driven sets are not adequate as continuous power supplies (several weeks' duration). Everything else being equal, the engine selected for a continuous duty application should be the one with the largest piston displacement, the lowest piston speed, and the lowest brake mean effective pressure or the average pressure on the piston over one complete engine cycle at rated output. If a small continuous-duty engine generator set is required, consideration might be given to a special lower speed unit.

4.2.7 *Construction and Controls.* The basic electrical components are the engine generator set and associated meters, controls, and switchgear. Most installations include a single generator set designed to serve either all the normal electrical needs of a building or a limited emergency circuit. Sometimes the system includes two or more generators of different types and sizes, serving different types of loads. Also, two or more generators may be operating in parallel to serve the same load.

4.2.8 *Typical Engine Generator Systems.* This information is designed to help in selecting the electrical components of a generator installation. No attempt has been made to cover every situation that might arise.

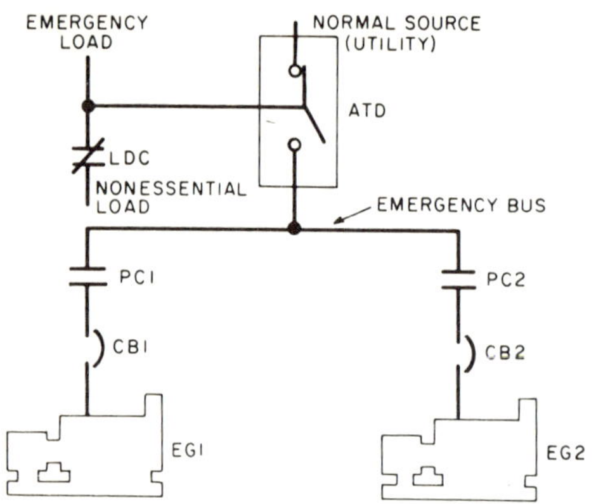

Fig 7
**Two Engine Generator Sets Operating
in Parallel**

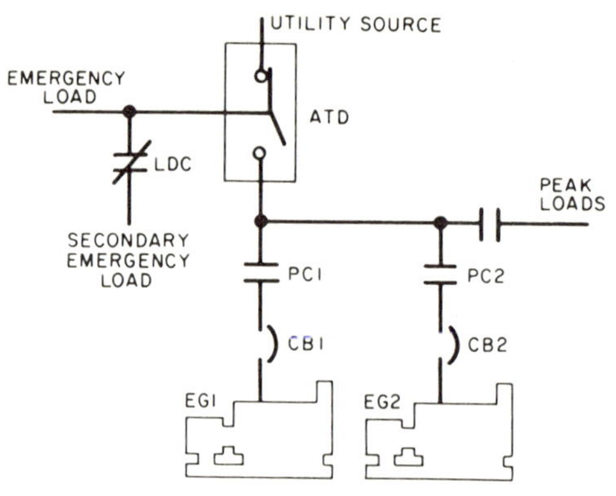

Fig 8
Peaking Power Control System

In Figs 7–11 the following abbreviations are used:

ATD = Automatic transfer device (automatic transfer switch or electrically operated circuit breaker)

CB = Circuit breaker

EG = Engine-driven generator set

LDC = Load-dumping contactor, electrically operated, mechanically held

PC = Paralleling contactor, electrically operated, mechanically held

Fig 7 shows a standby power system where if power fails from the normal source, both engines automatically start. The first generator to reach operating voltage and frequency will actuate load dumping circuits and cause the remaining load to transfer to this generator. When the second generator is in synchronism, it will be paralleled automatically with the first. After the generators are paralleled, all or part of the dumped load is reconnected if the standby capacity is adequate.

If one generator fails, it is immediately disconnected. A proportionate share of the load is dumped to reduce the remaining load to where the remaining generator can handle it. When the failed generator is reinstated, the dumped load is reconnected.

When the normal source is restored, the load is retransferred and the generators are automatically disconnected and shut down.

With the system shown in Fig 8, idle standby generator sets can perform a secondary function by helping to supply power for peak loads. Depending on the load requirements, this system starts one unit or more to feed peak loads while the utility service feeds the emergency loads. When the second generator is in synchronism, it will be paralleled automatically with the first. If the utility service fails, the peak loads are automatically disconnected and the generators pick up the emergency loads through the transfer device.

Fig 9 shows a standby power system where there is a split emergency load with one load being more critical than the other.

When the prime source fails, both generators start. If load 1 is the preferential load, the generator that reaches operating speed first is put on the line by automatic transfer device 3 to feed load 1 through automatic transfer device 1. When the other generator reaches operating speed, it then feeds load 2. If the generator feeding load 1 fails at

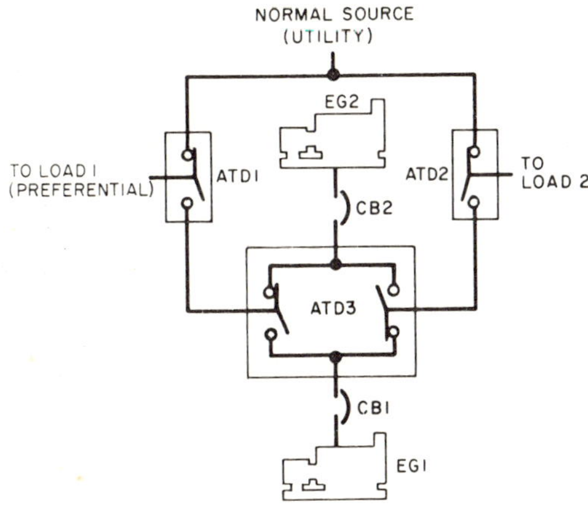

Fig 9
**Three-Source Priority Load Selection
System**

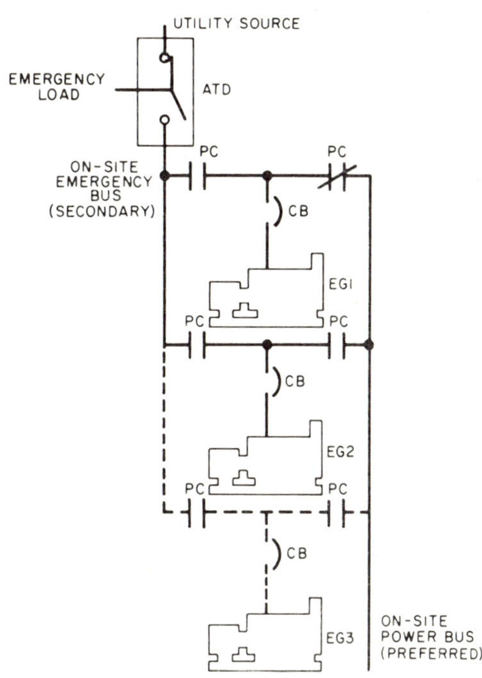

Fig 10
**Combination On-Site Power and
Emergency Transfer System**

any time, the other generator will be transferred from load 2 and take over load 1. When the prime source is restored, both loads are retransferred to the normal source and the generators shut down.

The system shown in Fig 10 provides switching and control of utility and on-site power. Two on-site buses are provided, (1) an on-site power bus (preferred) supplies continuous power for computer or other essential loads, and (2) an emergency bus (secondary) supplies on-site generator power to emergency loads through an automatic transfer device if the utility service fails.

In normal operation, one of the generators is selected to supply continuous power to the preferred bus (here EG1). Simplified semiautomatic synchronizing and paralleling controls permit any of the idle generators to be started and paralleled with the running generator to alternate generators without load interruption. Anticipatory failure circuits permit load transfer to a new generator without load interruption. However, if the generator enters a critical failure mode, transfer to a new generator is made automatically with load interruption.

Many loads such as lighting, fire alarms, heating, and air-conditioning are fed by the utility service through the transfer device. If the utility fails, idle generators are automatically started and assume these loads through the automatic transfer device.

Fig 11 is similar to Fig 8 where idle standby generator sets can perform a secondary function by helping to supply power for peak loads. Depending on the load requirements, this system starts one or both generators to feed the peak load by switching ATD2 while the utility service continues to supply the emergency and prime loads. The second generator is paralleled automatically with the first.

If the utility service fails, the emergency and prime loads are automat-

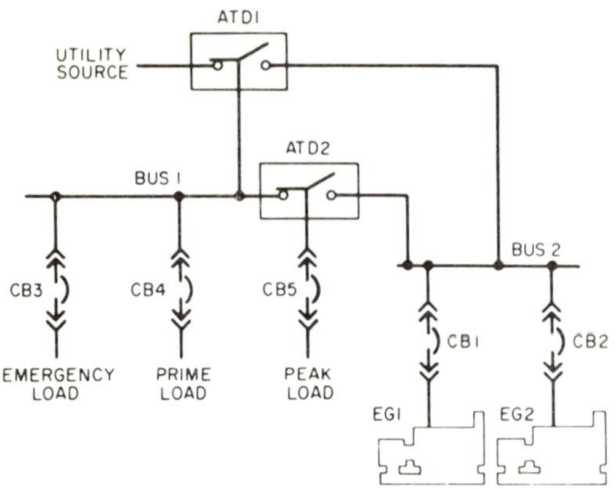

**Fig 11
Dual-Engine-Generator Standby System**

ically transferred to the emergency generators. Depending on generator capacity, the peaking load may be left on, or CB5 may be dropped for the emergency.

4.2.9 *Special Considerations.* Unusual conditions of altitude, ambient temperature, or ventilation may require either a larger generator to hold down winding temperatures or special insulation to withstand higher temperatures. Generators operating in the tropics are apt to encounter excessive moisture, high temperature, fungus, vermin, etc, and may require special tropical insulation and space heaters to keep the windings dry and the insulation from deteriorating.

4.2.10 *Engine Generator Set Rating.* For some buildings the maximum continuous generator load will be the total load when all equipment in the building is operating. For others it may be more practical and economical to set up an emergency circuit or circuits so that only certain essential lights and equipment, and perhaps just one elevator, can be operated when the load is on the standby generator.

4.2.11 *Motor-Starting Considerations.* Knowing the maximum mo-

mentary voltage dip that is acceptable in the circuit, it is possible to select the size of the engine-driven generator that will be able to start given size motors without exceeding the voltage dip. If it is possible that two motors may start together, the sum of their horsepower ratings must be used as a basis for estimating motor starting requirements or controls provided for separate starting.

Engines driving generators must be sized to handle the continuous kilowatt load to be supplied to the load plus motor starting requirements and the generator losses at the number of kilowatts selected.

In sizing the engine generators for motor starting, the "locked-rotor" or inrush kilovolt-ampere rating of the motors must be used. Manufacturers' data can usually be obtained, giving the maximum rating in kilovolt-amperes of the engine generator, as well as its continuous rating. The maximum rating would be the maximum number of short-duration kilovolt-amperes available for motor starting duty without exceeding a specified voltage dip. Motor starting load has a very low power factor which must be considered in calculating the voltage dip. One other important consideration is the effect of the generator voltage dip on the motor starting torque. The starting torque is proportional to the kilovolt-ampere input to the motor, but since a voltage dip to 70 percent of rated voltage results in a reduction of power to approximately 50 percent of the number of kilovolt-amperes into the stalled rotor, and thus a 50 percent reduction of motor starting torque, problems could result in starting motors under load, unless this factor is taken into consideration.

Generators are usually sized for the maximum continuous kilovolt-ampere demand. Should there be unusual high inertia loads to start without benefit of reduced voltage starting, or if voltage and frequency regulation other than specified cannot be tolerated during the startup period a larger generator may be required.

4.2.12 *Load Transient Considerations.* A voltage regulator with sufficient response is required to minimize voltage dips or rises after load transients. The engine generator set must be of sufficient size and design capability to minimize the effect of load transients. Many industrial applications can tolerate large voltage dips (up to 20 percent usually, but as much as 35 percent in special cases) as long as they are not so great as to cause motor contactors to drop out or automatic brakes to set. Solid-state controls and computers will be affected.

4.2.13 *Manual Systems.* Manually controlled standby service is the simplest and lowest cost arrangement and may be satisfactory where an attendant is on duty at all times and where automatic starting and transfer of the load is not a critical requirement.

4.2.14 *Automatic Systems.* In order for engine-driven generators to provide automatic emergency power, the system must also include automatic engine starting controls, automatic battery charger, and an automatic transfer device. In most applications, the utility source is the normal source and the engine generator set provides emergency power when utility power fails. The utility power supply is monitored and engine starting is automatically initiated once there is a failure or severe voltage or frequency reduction in the normal supply. It automatically transfers the load as soon as the standby generator stabilizes at rated voltage and speed. Upon restoration of normal supply, the transfer device automatically retransfers the load and initiates engine shutdown.

4.2.15 *Automatic Transfer Devices.* Transfer equipment for use with engine generator sets is similar to that used with multiple-utility systems, except for the addition of auxiliary contacts that close when the normal source fails. These auxiliary contacts initiate the starting and stopping of the engine-driven generator. The automatic transfer device may also include accessories for automatic exercising of the engine generator set and a 5 min unloaded running time before shutdown. For additional information on automatic transfer devices refer to Section 4.3.

4.2.16 *Engine Generator Set Reliability.* To keep the engine in good condition, whenever it starts it should run for a sufficiently long time so that all parts reach their normal operating temperatures. In case emergency power is called for only briefly, it is desirable that the engine continue to run for about 15 min after normal power has been restored. A programmed control should be added that starts the engine once a week and operates it for a set period of time, preferably under load.

Reliability and satisfaction of the standby power supply partially depends on the engine generator installations. Some key points to consider are the following:

(1) The water supply for the engine may be disrupted by the power interruption. An isolated water supply or a closed system should be specified.

(2) Antifreeze protection may be required. A room heater, electric immersion heater, or steam jacket improves starting reliability and solves the antifreeze problem at the same time. A loss-of-heat alarm may be required.

(3) A silencer on the intake and exhaust may be required by law and will deter complaints of noise.

(4) Fuel supply systems must meet local laws, regulations, and insurance re-

quirements. The capacity stored depends on available guaranteed quick delivery replacement, including Sundays and holidays and under all weather conditions.

(5) Fuel supplies stored underground are desirable. Above ground, antifreeze protection may be required.

(6) Gasoline and diesel fuels deteriorate if they stand unused for a period of several months. Normal testing and running on line may be used to keep fuel fresh. Inhibitors may be added to the fuel. Fuel used for several purposes will tend to be fresh.

(7) Engine vibration transferred to the building should be dampened by rubber pads or springs, and flexible couplings should be used to fuel lines.

(8) Starting aids are of three types, of which (c) is frequently preferred:

(a) Air heated before it reaches the cylinders

(b) Volatile starting fluid injected into the engine air

(c) Engine block maintained warm

4.2.17 *Air Supply and Exhaust.* Exhaust piping inside the building should be covered with gas-tight insulation to protect personnel and to reduce room temperature. The exhaust piping must be of sufficient diameter to avoid exhaust back pressure. Consideration should be given to dissipating the exhaust away from air intakes and with a minimum of air pollution.

Some means of providing free flow of fresh air into the generator room is necessary to keep the atmosphere comfortable for personnel and to make clean, cool air available to the engine.

4.2.18 *Noise Reduction.* Vibration must frequently be isolated from structures to reduce noise. Noise-reducing mufflers are rated according to their degree of silencing by such terms as "commercial," "moderate" or "semicritical," and "high degree" or "critical," and are usually required to meet noise standards. An intake muffler is not usually installed since an engine normally is supplied with at least one intake air cleaner that also serves as an intake silencer.

4.2.19 *Fuel Systems.* To simplify the fuel supply system, the fuel tank should be as close to the engine as possible. When gasoline or LP fuel is used, it normally cannot be stored in the same room with the engine because there is a danger of fire or fumes. However, when diesel fuel is used, it can generally be stored in the same room as the engine. The building code or fire insurance regulations should be checked to determine whether the fuel storage tank may be located beside the generator set, in an adjacent room, outside, or underground.

An engine equipped to operate on gasoline, LP gas, or diesel fuel stored in a nearby tank is a self-contained system that does not depend on outside services. It is dependable and affords independent standby protection.

4.2.20 *Governors and Regulation.* Governors are of two types, droop and isochronous. With a droop-type governor, the engine's speed is slightly higher at light loads than at heavy loads, while an isochronous governor maintains the same steady speed at any load up to full load:

$$\text{speed regulation} = \frac{\text{no-load r/min} - \text{full-load r/min}}{\text{full-load r/min}} \times 100\ \%$$

A typical speed regulation for a droop-type generator is 3 percent. Thus if speed and frequency at full load are 1800 r/min and 60 Hz, at no load they will be about 1854 r/min and 61.8 Hz. A droop-type governor usually is set so that it holds the desired nominal speed at full load.

Under steady load, frequency tends to vary slightly above and below the normal frequency setting of the governor. The extent of this variation is a measure of the stability of the governor. An isochronous governor should maintain frequency regulation within $\pm \frac{1}{4}$ percent.

When load is added or removed, speed and frequency dip or rise momentarily, usually 1 to 3 s, before the governor causes the engine to settle at a steady speed at the new load.

4.2.21 *Starting Methods.* Most engine generator sets utilize a battery-powered electric motor for starting the engine. The confidence level for dependable starting is no better than the reliability of the battery and its charger. A pneumatic or hydraulic system normally is used only where starting of the electric plant is initiated manually.

4.2.22 *Lighting and Battery Charging.* Some generator room lights may be powered by the engine batteries, especially if the set is to be started manually. In addition to the battery-charging generator on the set, a separate automatic battery charger is recommended for maintaining battery charge when the generator is not running. The input of the automatic battery charger should be connected to the load side of the transfer switch.

4.2.23 *Additional Information.* Engine generator specifications are treated in greater detail in EGSMA GTD2-1971, Glossary of Standard Industry Terminology and Definitions, and EGSMA EGS1-1970, Standard Specifications for Standby Engine Driven Generator Sets.

4.3 Multiple Utility Services

4.3.1 *Introduction.* Multiple utility services may be used as an emergency or standby source of power. Required is an additional utility service from a separate source and the required switching equipment. Fig 12 shows automatic transfer between two low-voltage utility supplies. Utility source 1 is the normal power line and utility source 2 is a separate utility supply providing emergency power. Both circuit breakers are normally closed. The load must be able to tolerate the few cycles of interruption while the automatic transfer device operates.

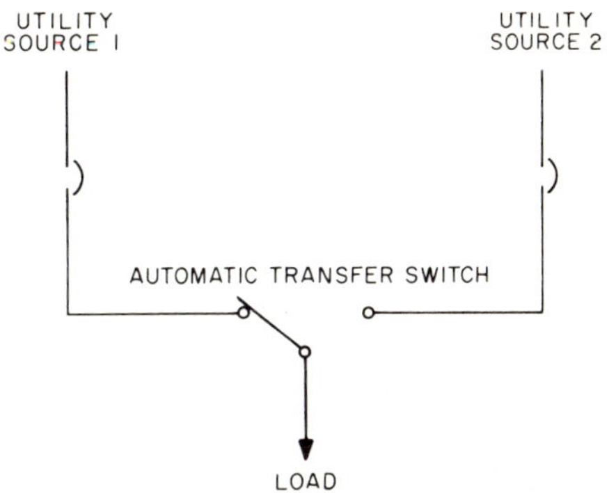

Fig 12
Two-Utility-Source System Using One Automatic Transfer Switch

4.3.2 *Closed-Transition Transfer.* If the utility will permit the two sources of supply to be connected together momentarily, the transfer device should be provided with controls for both open (normal supply opened before the emergency supply is closed) and closed (emergency supply closed before the normal supply is opened) transition. With closed transition, the utility can notify the customer to transfer to the emergency source in order to take the normal supply out of service for maintenance and repair without the momentary interruption which occurs with open transition.

4.3.3 *Utility Services Separation.* Use of multiple utility service is economically feasible when the local utility can provide two or more service connections over separate lines and from separate supply points that are not apt to be jointly affected by system disturbances, storms, or other hazards. It has the advantage of relatively fast transfer in that there is no 5 to 15 s delay as there is when starting a standby engine generator set. A separate utility supply for an emergency should not be relied upon unless total loss of power can be tolerated on rare occasions.

A no-potential alarm should be installed on the emergency supply so that the utility can be notified and emergency precautions taken if the emergency supply is lost. Additional reliability has been obtained in rare instances where the services are available from different utility companies.

4.3.4 *Simple Automatic Transfer Schemes.* Automatic switching equipment may consist of two circuit breakers interlocked as shown in Fig 13. Circuit breakers are generally used for pirmary switching where the voltage exceeds 600 V. They are more expensive but are safer to operate and the use of fuses for overcurrent protection is avoided. Relaying is provided to transfer the load automatically to either source if the other one fails, provided that circuit is energized. The supplying utility will normally designate which source is for normal use and which for emergency. If either supply is not able to carry the entire load, provisions must be made to drop noncritical loads before transfer takes place. If load can be taken from both services, the two R circuit breakers are closed and the tie circuit breaker is open. The three circuit breakers are interlocked to permit any two to be closed but prevent all three from being closed. This arrangement has the advantage that the momentary transfer outage will occur only on the load supplied from the circuit which is lost. However, the supplying utility may not allow load to be taken from both sources, especially since a more ex-

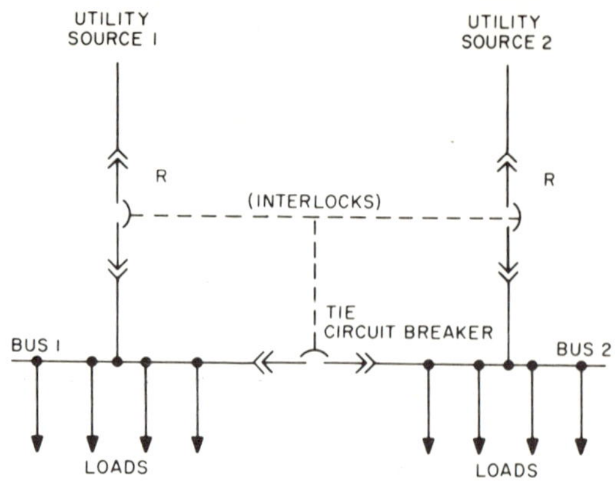

Fig 13
Two-Utility-Source System where any Two Circuit Breakers Can Be Closed

pensive totalizing meter may be required. A manual override of the interlock system should be provided so that a closed transition transfer can be made if the supplying utility wants to take either line out of service for maintenance or repair and a momentary tie is permitted.

If the supplying utility will not permit power to be aken from both sources, the control system must be arranged so that the circuit breaker on the normal source is closed, the tie circuit breaker is closed, and the emergency source circuit breaker is open. If the utility will not permit dual or totalized metering, the two sources must be connected together to provide a common metering point and then connected to the distribution switchboard. In this case the tie circuit breaker can be eliminated and the two circuit breakers act as a transfer device. Under these conditions the extra cost of circuit breakers can rarely be justified.

The arrangement shown in Fig 13 only provides protection against failure of the normal utility service. Continuity of power to critical loads can also be disrupted by (1) an open circuit within the building (load side of the incoming ser-

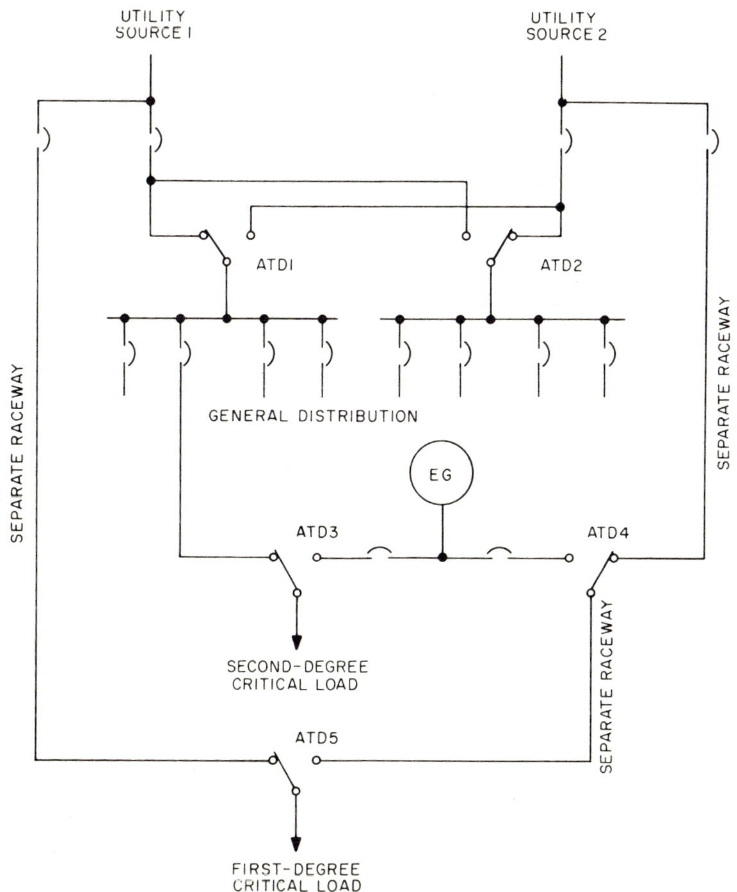

Fig 14
**Two Utility Sources Combined with an Engine Generator Set
to Provide Varying Degrees of Emergency Power**

vice), (2) an overload or fault tripping out a circuit, or (3) electrical or mechanical failure of the electric power distribution system within the building. It may be desirable to locate transfer devices close to the load and have the operation of the transfer devices independent of overcurrent protection. Multiple transfer devices of lower current rating, each supplying a part of the load, may be used rather than one transfer device for the entire load.

Reliability of multiple utility service systems can be improved by adding a standby engine generator set capable of supplying the more critical load. Such an arrangement, using multiple automatic transfer switches, is shown in Fig 14.

4.3.5 *Overcurrent Protection.* Caution should be exercised to assure that transfer control and operation do not in any way detract from overcurrent protection and vice versa. Furthermore, the transfer and overcurrent protective devices should be so arranged that means for disconnecting incoming service is conventional and readily accessible.

4.3.6 *Transfer Device Ratings and Accessories.* The required characteristics of transfer devices should include

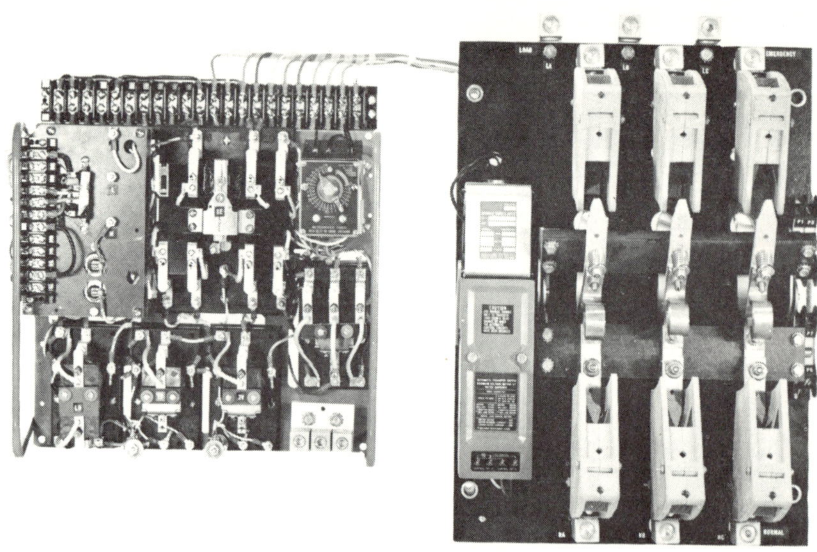

**Fig 15
Modular Type Automatic Transfer Switch Suitable for
All Classes of Load**

their capabilities to (1) close against inrush currents without contact welding, (2) carry full rated current continuously without overheating, (3) withstand available short-circuit currents without contact separation, and (4) properly interrupt the loads to avoid flashover between the two utility services.

In addition to considering each of the above individually, it is also necessary to consider the effect each has on the other. Particular consideration must be given to coordination between automatic transfer switches and overcurrent protection. High fault currents create electromagnetic forces within the contact structure of circuit breakers which help provide fast opening and therefore minimum clearing time. However, automatic transfer switches designed to withstand high fault current utilize these electromagnetic forces in a reverse manner to assure that the transfer switch contacts remain closed until the fault has been cleared. For these reasons, transfer switching devices should be selected from those designed and approved for the purpose.

Load transfer devices are available in the following forms:

(1) Automatic transfer switches are available in ratings from 30 to 3000 A, 600 V through 15 kV (Fig 15) [2].

(2) Automatic power circuit breakers consist of two or more circuit breakers which are mechanically or electrically interlocked, or both.

(3) Manual transfer switches (600 V) are available in current ratings from 30 to 200 A.

(4) Manual or electrically operated bolted pressure switches (600 V), fusible or nonfusible, are available from 800 to 6000 A.

Automatic transfer switches are used with manual reset from "emergency" to "normal" position. Features and accessories to be considered include the following:

(1) Relays factory set for 95 percent pickup and 90 percent dropout (adjustable) with a time delay of 1 to 10 min af-

ter pickup will provide adequate voltage sensing for most industrial plant loads. Lower dropout may be necessary if significant voltage drop is produced by starting large motors.

(2) Test switch to simulate a power failure to provide a periodic test of the emergency source and the transfer operation.

(3) A potential alarm on the emergency source. If the emergency supply is protected by fuses ahead of the transfer switch, either on the utility system or in the service equipment, indicators should be provided on all three phases to detect a blown fuse. This will require two or three potential transformers on the emergency source instead of the one normally supplied.

(4) Controls for closed transition if the utility will permit the two sources to be momentarily tied together and the switch design permits. This will allow manual transfer from the normal to the emergency feeder and back without the momentary transfer interruption.

Consideration should also be given to minimum voltage at which load will operate satisfactorily. The time of interruption that the load can tolerate will determine the maximum time delay permitted on the load supply devices to override momentary interruptions and if critical, the type of transfer device.

4.3.7 *Voltage Tolerances*. The basic standard for the voltage tolerance for utilization equipment is ANSI C84.1-1970, Voltage Ratings for Electric Power Systems and Equipment (60 Hz). These limits are based on the T frame motor with slightly reduced limits for equipment other than motors. In the case of special equipment, the manufacturer's tolerance limits should be obtained. Care should be taken in determining the minimum voltage for transfer switch operation to distinguish between equipment which is not damaged

by low voltage, even though operation is unsatisfactory, such as incandescent lamps and resistance heaters, and equipment which will be damaged or cause damage or unsafe conditions by low voltage.

4.3.8 *Transferring Motor Loads*. Transferring motor loads between two sources requires special consideration. Although the two sources may be synchronized, the motor will tend to slow down upon loss of power and during transfer, thus causing the residual voltage of the motor to be out of phase with the oncoming source. The speed of transfer, total inertia, and motor and system characteristics are involved. On transfer, the vector difference could cause serious damage to the motor and the excessive current drawn by the motor may trip the overcurrent protective device. Both motor loads with relatively low load inertia in relation to torque requirements such as pumps and compressors, and large inertia loads such as induced draft fans, etc, that keep turning near synchronous speed after loss of power for longer time, are subject to the hazard of out-of-phase switching. Automatic transfer switches can be provided with accessory controls that initiate motor disconnect prior to transfer and reconnect the motor after transfer and when the residual voltage has been substantially reduced.

4.3.9 *Operation of a Typical System*. Fig 16 illustrates a practical system supplying electric power to a manufacturing plant. The system is designed for initial operation with utility line 1 utilized as the normal source and utility line 2 utilized as a normally open auxiliary source. The two utility lines are synchronized with each other so that they can be paralleled, but are not normally operated in this manner. However, unless the relaying is designed for

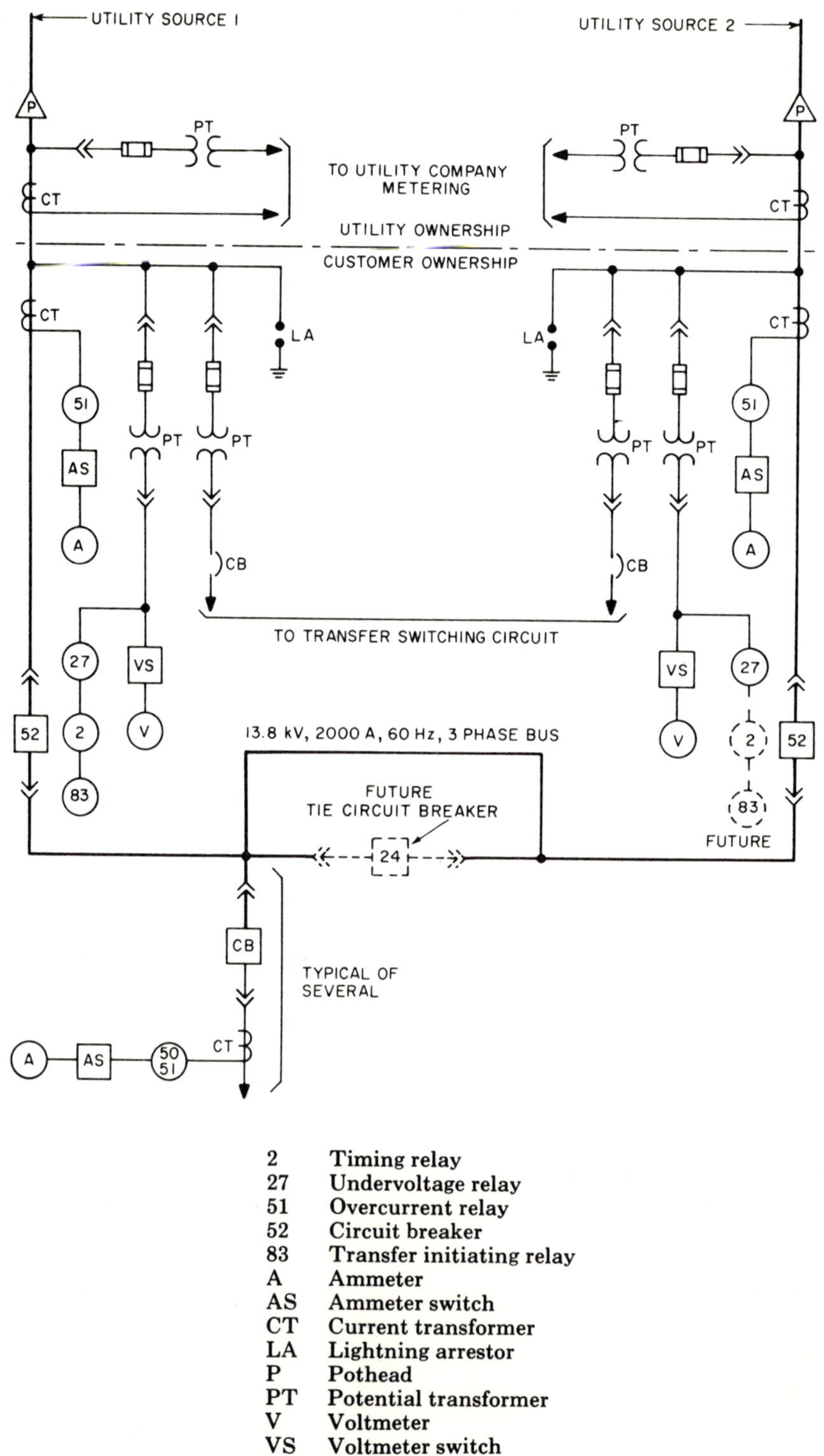

2	Timing relay
27	Undervoltage relay
51	Overcurrent relay
52	Circuit breaker
83	Transfer initiating relay
A	Ammeter
AS	Ammeter switch
CT	Current transformer
LA	Lightning arrestor
P	Pothead
PT	Potential transformer
V	Voltmeter
VS	Voltmeter switch

**Fig 16
Typical System Supplying Electric Power to Manufacturing Plant**

it, the operation of the two incoming lines in parallel must be kept to a minimum, that is, the switching time.

Utility lines 1 and 2 enter the plant from a substation some distance away through underground conduits separated by about 1 ft and encased in the center of a 3 by 3 ft concrete enclosure 7 ft below the surface for protection.

Operation is as follows.

(1) If voltage on the normal source (line 1) drops to 65 percent for several cycles, the undervoltage relay will deactivate, trip the circuit breaker in line 1, and close the circuit breaker in line 2 (if acceptable voltage is present on line 2).

(2) When voltage is restored to line 1, the undervoltage relay is activated and initiates a timer. After the voltage has been present on line 1 for a predetermined time (usually 1 to 10 min), the circuit breaker in line 1 closes, after which the circuit breaker in line 2 opens.

(3) If there is no voltage present on line 2 when line 1 loses voltage, the circuit breaker in line 2 will not close. When voltage is restored to line 1 the circuit breaker in line 1 will immediately close.

(4) If a fault or overload occurs on the load side of either incoming circuit breaker, a lockout relay keeps both circuit breakers open, disabling the automatic transfer system until manually reset.

As power demands increase this system can be expanded by inserting a tie circuit breaker in the 13.8 kV bus and additional relaying. Part of the load would then be supplied by each utility line with transfer of the entire load to the live feeder should power be lost to one line. Load shedding of noncritical loads could be incorporated if necessary.

Operation would then be as follows:

(1) A loss of voltage or a decrease of voltage below 65 percent on either utility line will cause that normally closed circuit breaker to open and the normally open tie circuit breaker to close. When normal voltage returns, the open utility line circuit breaker will close in a preset time (1 to 10 min) after which the tie circuit breaker will open.

(2) A simultaneous loss of voltage on both utility lines will cause both normally closed utility line circuit breakers to open and the normally open tie circuit breaker to close. A return to normal voltage on either utility line will cause that utility line circuit breaker to close in its preset time and the tie circuit breaker to remain closed. When normal voltage is established on the second utility line, that utility line circuit breaker will close in its preset time after which the tie circuit breaker will open.

(3) Fault current or overload current causing either utility line circuit breaker to open will also make the automatic closing feature of the tie circuit breaker inoperative until manually reset.

(4) If the utility company needs to take one line out of service, they notify the customer who then manually closes the tie circuit breaker and opens the line to be affected.

4.3.10. Conclusion. Approximately 90 percent of transfer schemes designed to switch from the prime power source to the emergency source for commercial installations utilize conventional double-throw transfer switches. For maximum system reliability the transfer switches are usually located close to the load rather than at the incoming prime power source. The use of interlocked service entrance circuit breakers for transfer schemes is usually limited to medium-voltage primary switching.

Two parallel in-phase separate utility sources on line continuously with proper relaying on a normally closed tie

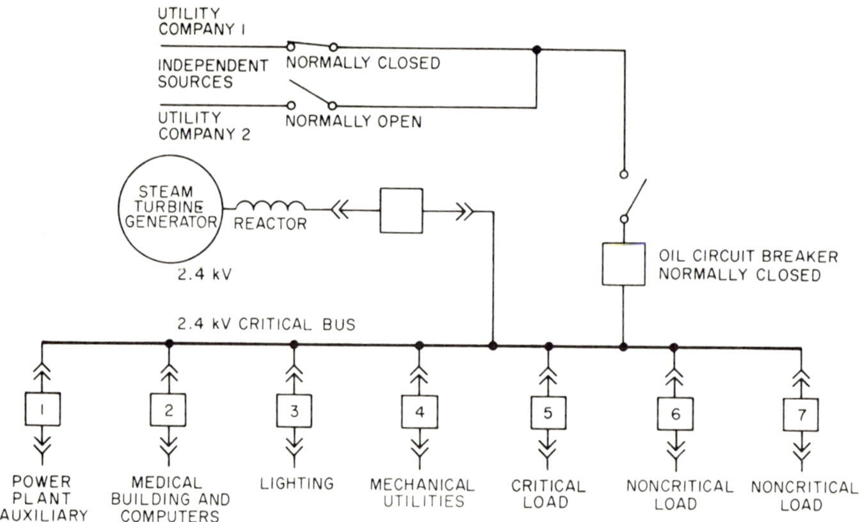

**Fig 17
Emergency and Standby Power System Using Steam-Turbine and
Dual Utility Supply**

circuit breaker provide increased reliability of electric power in some applications. This possibility should be investigated with the utility company representatives.

4.4 Turbine-Driven Generators

4.4.1 Introduction. Two general types of turbine prime movers for electrical generators are available, steam and gas or oil.

4.4.2 Steam Turbine Generators. Steam is usually not available if all electric power has been lost, although there are independent steam supply systems which themselves may have uninterruptible electrical systems. In this case steam might be considered. There are compact steam turbines which could bring power onto the line in about 5 min. This is a rather special source of supply and details are not presented.

Steam turbines are used to drive generators which are larger than those which can be driven by diesel engines. However, steam turbines are designed for continuous operation and require a boiler with fuel supply and a source of condensing water. Thus they are expensive for use as an emergency or standby power supply and may have environmental problems involving fuel supply, noise, combustion product output, and heating of the condensing water.

Fig 17 shows an on-line steam turbine generator supplying a critical bus in parallel with one source of utility power. An alternate utility source can be manually switched on in a minute or so should there be a failure of the on-line utility source. The normal utility supply system should be large enough to supply the entire critical bus if the turbine is off.

Reverse current relays should be used to immediately separate the utility supply and the steam-driven generator so that emergency power is available on the critical bus. Any protective relay scheme should be coordinated to prevent backfeed to the utility service. At the same time two noncritical loads are shed so as not to overload the steam turbine. If the utility supply has failed, a manual transfer is made to the alternate source and the shed loads are reenergized. If the turbine has dropped from the bus, a decision is made to accept a new demand peak from the utility company or to wait for the return of energy from the turbine generator before energizing the two noncritical loads. Necessary protective relaying has not been shown in the figure (see IEEE Std 357-1973, Protective Relaying of Utility–Consumer Interconnections).

4.4.3 *Gas and Oil Turbine Generators.* The most common turbine-driven electric generator units employed for emergency or standby power today use gas or oil for fuel. Various grades of oil and both natural and propane gas may be used. Other less common sources of fuel are kerosene or gasoline. Service can be restored from about a 40 s minimum to several minutes for larger combustion turbine units.

Aircraft-type turbines driving generators have been commonly used where electric power may be needed for a few hours to days. Small industrial units have been developed. A small fuel storage facility for safety may be adequate, provided plans have been made by which additional fuel will be delivered when needed. Care should be taken to assure an adequate gas supply should this be the source of energy, since an uninterruptible supply from a public utility company may be very expensive or unavailable. An interruptible supply may not be available when needed during cold weather. Earthquakes may destroy extensive underground utility distribution systems, but local storage may be intact. Environmental considerations may require the use of low-sulfur oil which may be difficult to obtain.

Savings are possible by the installation of a turbine for emergency and standby power when used as a peak clipping unit to reduce the demand charge. This has the operating advantage of checking the unit often enough under load so that the operating personnel are familiar with the equipment and the unit is known to be ready for an emergency. See case histories in Section 5.

Turbine designs fall between two extreme categories. Aircraft-type engines embody highly sophisticated techniques for very light weight in relation to horsepower ($\frac{1}{4}$ to $\frac{1}{2}$ lb/hp). This shortens life. Fig 18 shows a typical generator of this type with a riser diagram of typical loads served. Fig 19 shows a modular packaged gas turbine. The opposite philosophy employs the massive, bulky design techniques of steam turbines in an effort to assure long life, but with 10 lb/hp. Both have their place depending on the cost justification and hours of usage per year. Reliable industrial designs are available within these ranges.

Accessory equipment for a gas turbine such as filtering, silencing apparatus, and vibration mounts may be required when climate conditions indicate contaminated dust-laden air or areas where noise and vibration level attenuation is required. Bands of noise from 75 to 9600 Hz are common and attenuation from 5 to 60 dB or more may be necessary in various bands.

A complete 750 kW gas turbine generator unit weighs approximately 13 000 lb and occupies less then 80 ft^2 of floor area. This factor of compact size and light weight can considerably reduce building costs and allow more efficient economic utilization of building space. Rooftop installations are both feasible and practicable.

The combustion turbine generator set exhibits superior performance regarding frequency control, voltage regulation, transient response, and behavior

(a)

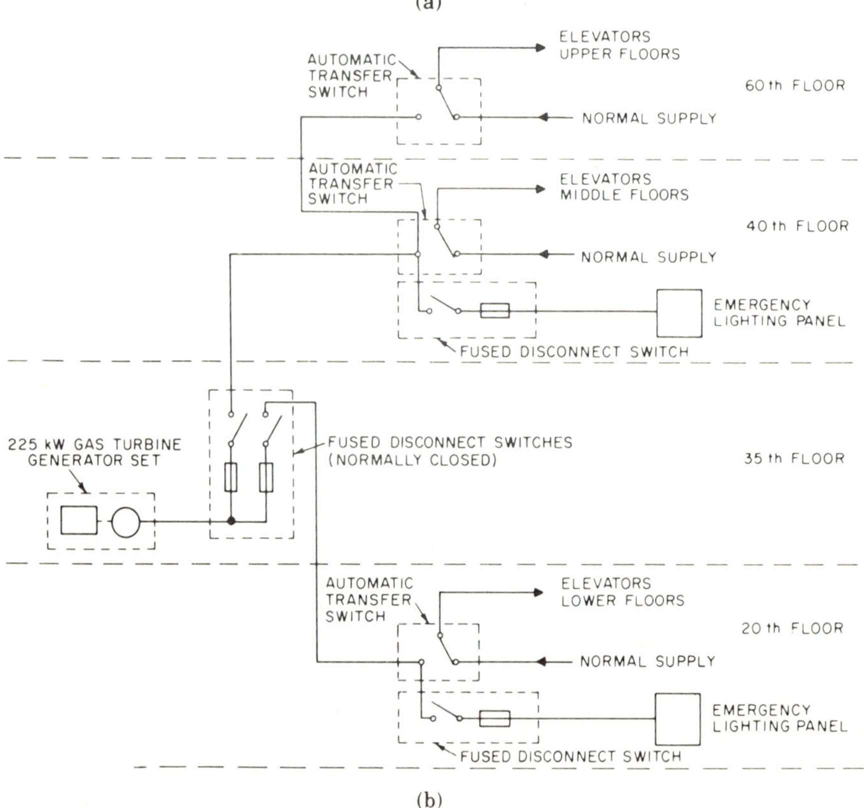

(b)

**Fig 18
Typical Gas-Turbine Generator and Riser Diagram of
Auxiliary Power System**

Fig 19
Modular Packaged Gas-Turbine Generator Set Mounted on Trailer

when operated in parallel with units of the utility supply.

Combustion turbine starting and loading is accomplished either automatically or manually. Warmup is not necessary.

There are four basic starting systems available for turbines:

(1) Electric motor supplied from batteries

(2) A small steam turbine

(3) A compressed air or gas system

(4) A small diesel engine

The controls for multiple-unit installations generally involve some interconnection considerations in a master control panel or synchronizing panel. The turbines may be programmed for automatic or manual operation, start sequence, and synchronizing if desired. Special protection such as differential relays may be required.

Fig 20 shows the reduction in output capacity as gas- or oil-fired combustion turbines are installed at increasing altitudes. High air input temperatures with the associated lower air density and low barometric pressures also reduce the available output. These limitations must be taken into account or the anticipated reliability and capacity may not be obtained.

Fuel consumption at sea level will be about 14 200 to 16 250 Btu/kWh output, depending upon size and variable factors. Thermal efficiency may be raised considerably if waste heat can be used.

Installed costs will run in the vicinity of $150 per kilowatt in the medium sizes from 800 to 6000 kW, including fuel tank and turbine generator package with switchgear, set in place, piped, and ducted.

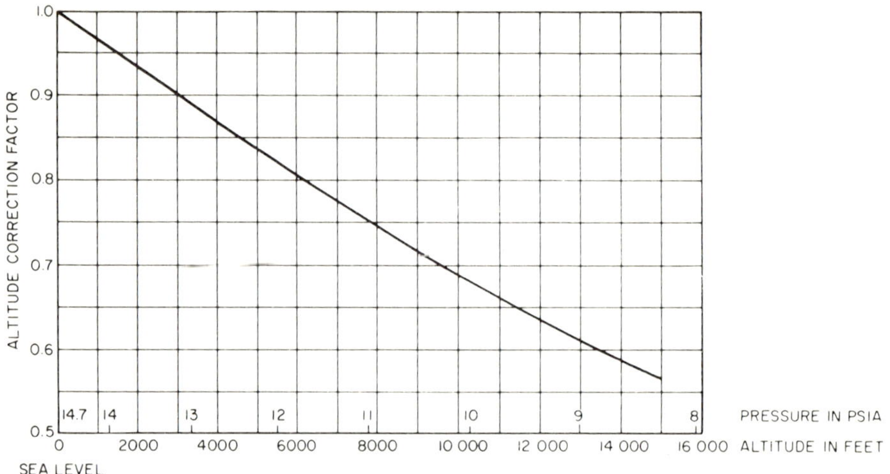

Fig 20
Typical Performance Correction Factor for Altitude

4.5 Mobile Equipment

4.5.1 *Introduction.* Perhaps one of the most overlooked sources of emergency or standby power is mobile equipment. For most industrial applications mobile equipment will include only two types, engine-driven and gas turbine-driven generators.

4.5.2 *Gas Turbine Generators.* Below 200 kW the economics of a gas turbine make difficult comparison with reciprocating engine-driven generators. However, above this range the gas turbine becomes more easily justified. The gas turbine has several advantages over the reciprocating engine type, a few of which are listed below:

(1) Low initial cost (in larger sizes)
(2) Black plant starting ability
(3) Easy to silence
(4) Not dependent on cooling water
(5) Relatively small and compact
(6) Light weight
(7) Low vibration
(8) Easy mobility
(9) Low air pollution
(10) Operates on a variety of fuels

4.5.3 *Available Sizes.* The range of sizes available for mobile generators starts at 10 kW and goes as high as 2700 kW for gas turbine units and from less than 1 kW up to 450 kW for engine-driven units. Larger capacities may be obtained by parallel operation. Fig 21 shows a typical mobile turbine generator unit with associated switchgear.

4.5.4 *Accessories.* Mobile generators come anywhere from a stripped down unit with nothing but the prime mover and generator to units complete with soundproof chamber, control panel, relaying, switchgear, intake and exhaust silencers, fuel tank, battery, and other required operating and safety devices.

4.5.5 *Mobile Equipment Rental.* Many times a user cannot justify the cost of purchasing a mobile generating unit for his use, but this problem should not eliminate the subject of mobile generators. Table 11 shows approximate costs of purchase and rental for various sizes of reciprocating engine-driven generating units.

There are three major sources of rental equipment.

(1) Many of the large engine generator sales companies have rental ma-

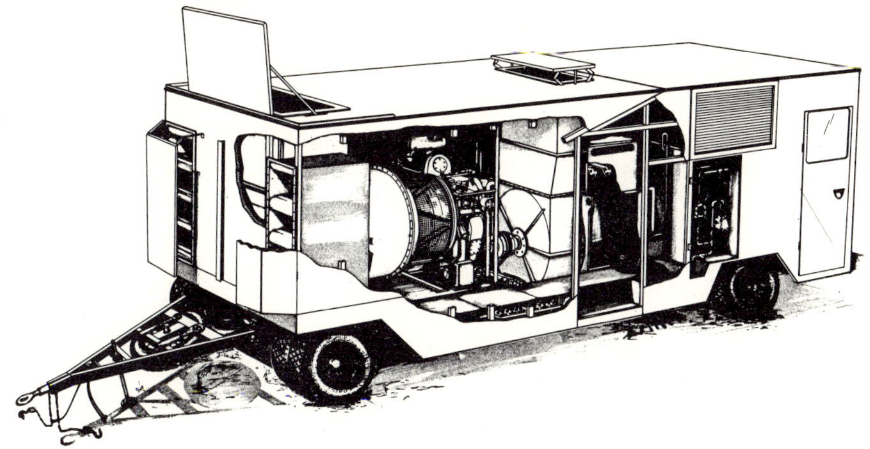

Fig 21
Typical 1200 kW Mobile Turbine-Driven Generator [12].

Table 11
Comparative Costs for Engine-Driven Generators
1971 Prices

Size (kW)	Purchase Price per Kilowatt (dollars)	Rental Cost per Kilowatt per Month (3 Shifts) (dollars)	Rental Cost per Kilowatt per 3-Shift Day (dollars)
135	66 – 89	10.60 – 14.20	1.19 – 1.58
190	75 – 110	11.90 – 16.00	1.33 – 1.78
250	67 – 90	10.70 – 14.40	1.20 – 1.60
335	70 – 94	11.10 – 14.90	1.23 – 1.64
400*	82 – 110	13.10 – 17.50	1.41 – 1.88

*The higher costs per kilowatt for the 400 kW unit show the rise in cost if a slower speed machine is used.

chines which are normally available in case of a local prolonged power outage.

(2) Some utility companies have mobile equipment which is made available to certain customers on an emergency basis.

(3) Most large contractors have at least one mobile unit that is used for temporary power at new job sites, and under certain conditions would be available for rental.

4.5.6 *Cooperative Purchase of Mobile Equipment.* Another way to avoid the high cost of purchasing mobile equipment is for several users to contribute and purchase one unit to provide standby power for any one of the users as it is needed. The drawback to relying on a rental unit or a cooperatively purchased unit is that in case of a widespread prolonged interruption, the generators usually go on a first-come first-served basis.

4.5.7 *Use of Utility Company Transformers.* Another type of mobile equipment is utility-owned transformers which are mounted on a truck. Most utility companies have these mobile transformers for maintaining power to an area while performing maintenance on the transformer normally supplying the area. These mobile transformers may be available to a user if one of his transformers fails.

Sometimes power is lost from the normal utility supply source, but a lower or higher voltage is available close by. A mobile transformer can be used for emergency power secured in time to prevent damage from freezing or in a quantity sufficient for partial production.

4.5.8 *Fuel Systems.* Growing shortages of gas and liquid fuel sources require advanced planning for a firm, readily available, dependable source of supply under any emergency condition foreseen. When natural gas is planned, a firm source should be available. A back-up source of propane should be considered. Where liquid fuels are planned, readily available storage at the point of usage should be considered. Since liquid fuels deteriorate during long storage, a system of use and replacement to assure a fresh supply must be planned.

4.5.9 *Agricultural Applications.* Farm standby tractor-driven generators are available, practical, and usually reasonable in cost since the prime mover is serving a dual purpose and is usually on hand for other uses (see EGSMA Std TDGS1-1972, Standard Specifications for Tractor Driven Generator Sets, and EGSMA IMFS1-1974, Standards and Recommendations for Installation and Maintenance of Farm Standby Electric Power).

4.6 Mechanical Stored Energy Systems

4.6.1 *Introduction.* Systems under this classification deliver uninterruptible power by converting kinetic energy (KE) contained in a rotating mass to electric energy:

$$\text{KE} = \frac{(WK^2)\,(\text{r/min})^2}{3.23 \times 10^6} \ \text{hp·s}$$

where W is the weight in pounds and K the radius of gyration in feet.

These systems provide an excellent "buffer" between the prime power source and loads which will not tolerate transients in voltage and frequency.

By switching to standby power systems during the power support time provided by these systems an uninterruptible power supply system may be secured for any length of time.

4.6.2 *Typical System Types.* Many practical systems are in use. Six systems will be described. Other configurations can be assembled, limited only by the need, economics, and ingenuity of the designer.

(1) System 1 (Fig 22) is composed of an induction motor with low-slip characteristics which drives a flywheel and a synchronous generator. Under full rated load the output frequency is 59.8 Hz. When input power is lost, the energy stored in the flywheel drives the generator. The frequency is maintained above 59.5 Hz for time intervals of up to 0.5 s. The time interval for which frequency can be maintained is proportional to the ratio of the flywheel inertia to the load, for any given operational speed. To keep the weight of the system low, high speed is desirable. To keep noise output down and to optimize reliability, low speed is desirable. Commonly the system is operated at 1800 r/min as a good compromise speed.

Most interruptions of alternating-current input power are of short duration, lasting for intervals of from ½ to 30 cycles. Hence this system will minimize loss of boiler controls and computer time from line transients at relatively low cost. However, its operational speed and hence its output frequency are directly in proportion to input frequency. If a backup engine generator set is used in emergencies to power this type of critical load protection, it should have a highly accurate frequency control which will hold 60 ± 0.25 Hz for all load conditions. To do this it is necessary to provide a separate engine gener-

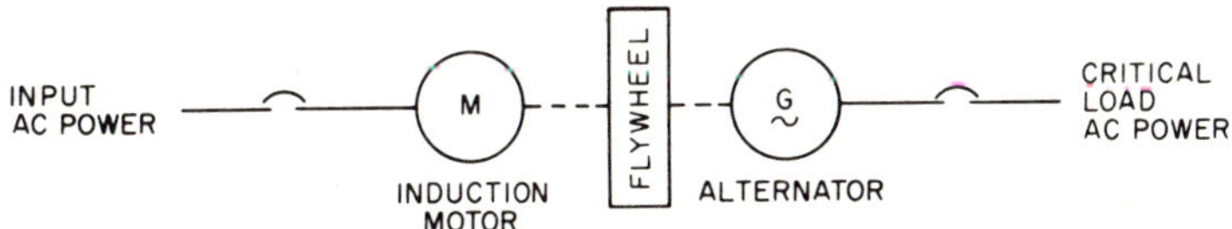

Fig 22
Simple Inertia-Driven "Ride-Through" System

ator set for use only for backup to computer power in order to prevent frequency transients imposed by other loads turning on or off, such as air-conditioning, elevators, etc. This backup generator is connected to the induction motor. The engine generator set must be oversized by approximately 2 times steady-state load in order to start the low-slip induction motor.

Upon loss of input power, the decay of output frequency is essentially linear down to 85 percent of speed while the output voltage can be held to ±3 percent during this period.

For units up to 100 kVA a minimum of 300 ms ride-through is required for a system whose minimum tolerable input frequency is 59.5 Hz. For a system which can tolerate an input frequency of 58.5 Hz, the ride-through capability would be 1.8 s.

System 1 has relatively low cost but provides minimum protection of the computer against alternating-current input power loss or transients. Medium-size units (125 kVA) cost about $200 per kilovolt-ampere.

An option available for these units senses the output frequency of the generator to initiate the shutdown sequence. This option effectively measures the energy stored in the flywheel, regardless of the sequence of voltage dips and momentary interruptions which may have reduced the speed of rotation of the set. Thus the assurance is greater that the shutdown initiating signal is given at the optimum time,

than is possible with input undervoltage or voltage-loss sensing and a time delay.

(2) System 2 (Figs 23 and 24) is effectively an alternating-current flywheel motor generator set with a direct-current machine and a battery bank added. In normal operation, the alternating-current motor drives the alternating current generator to supply the load. An option is available which permits the direct-current machine to act as a generator to trickle charge the batteries. Upon loss of alternating-current power, the alternating-current starter drops out and a direct-current contactor closes, applying battery voltage to the direct-current machine. The direct-current machine then operates as a motor to drive the generator. The inertia of the flywheel and the rotating machines buffer the transitions between normal and emergency operation.

When alternating-current input power is restored, the system reverts automatically to the normal mode, and the direct-current machine recharges the batteries (if that option is included). The length of interruption through which the load can be sustained is determined by the amount of installed battery capacity.

The battery capacity may be chosen based upon the time required to start and bring on line an auxiliary generator, or an orderly shutdown may be employed.

(3) System 3 (Fig 25) consists of an induction motor which drives a fly-

**Fig 23
Battery-Supported Inertia System**

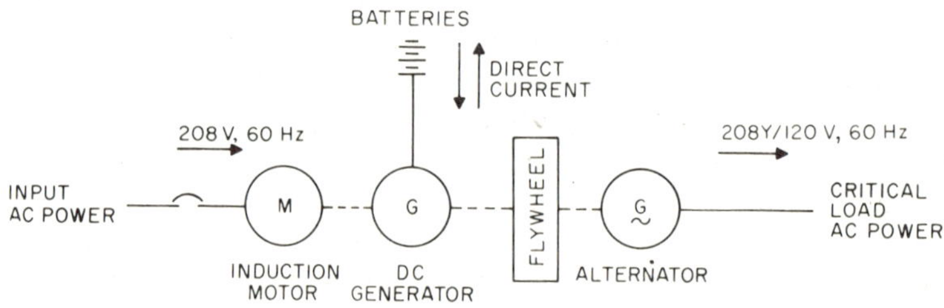

**Fig 24
Diagram of Battery-Supported Inertia System**

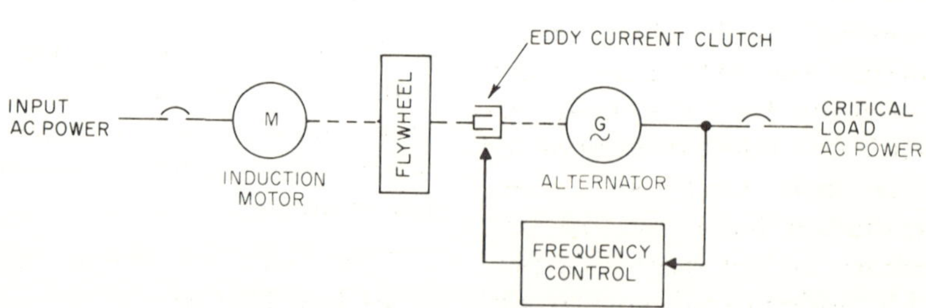

**Fig 25
Constant-Frequency Inertia System**

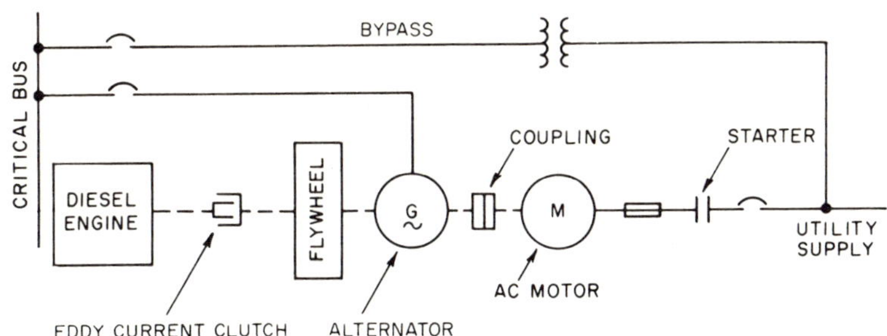

**Fig 26
Rotating Flywheel No-Break System**

wheel and an eddy-current clutch at a fixed speed. The generator operates at a lower speed than the flywheel by controlling the slip of the eddy-current clutch, and output frequency is maintained at 60 Hz ± 0.25 Hz by this control

On loss of alternating-current input power the generator receives energy stored in the flywheel. As the flywheel slows, the slip of the eddy-current clutch is reduced so as to maintain the 60 Hz output.

Power to the critical load is maintained for up to 15 s after loss of alternating-current input power. This provides time to start and switch in the backup power source which is normally an engine generator set with adequate power capacity to start and operate the connected loads.

Efficiencies are poor, being usually less than 55 percent at full load. Reliability is generally limited by the bearings supporting the extremely heavy and high-speed flywheel. Purchase cost is about $400 per kilovolt-ampere for a medium-size unit (125 kVA), not including installation, based on 1970 cost. For preliminary budget figures 50 percent should be added to this cost to cover the installation.

(4) System 4 is shown in Fig 26. An induction motor is driven from the utility supply and this motor is directly coupled to an alternator with its own excitation and voltage-regulating system. Coupled directly to the motor generator set is a large flywheel with one member of a magnetic clutch attached to the flywheel. The other half of the clutch is connected to a diesel engine or other prime power. During the transient period of power changeover, the kinetic energy in the flywheel is used to generate power and the voltage regulator maintains the voltage. With proper selection of components to minimize the start and runup times of the diesel engine, the frequency dip can be kept to approximately 1.5 to 2 Hz without paying a premium. Thus with a steady-state frequency of 59.5 Hz, the transient frequency would be from 57.5 to 58 Hz. The time for the diesel to start, come up to speed, and take load would normally be from 6 to 12s.

The cost of the equipment is from $700 to $800 per kilowatt with 150 kW being the practical limit, without sacrifice of performance or increase of cost per kilowatt.

The advantages of system 4 are

(a) Low first cost

(b) Moderate maintenance cost

(c) No heavy battery or battery ventilation equipment required

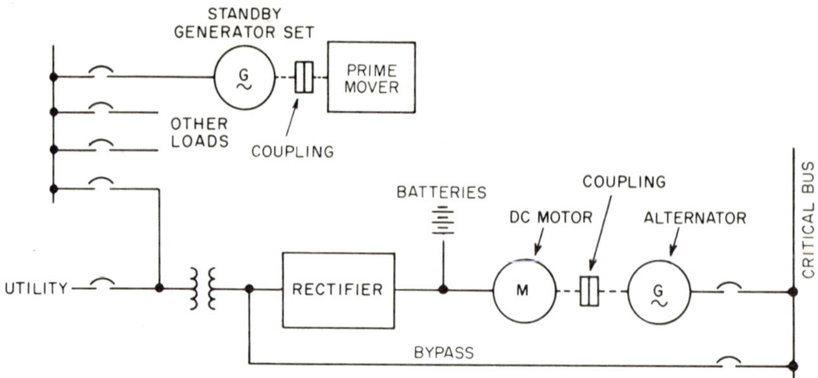

Fig 27
Engine-Generator Supported Battery Inertia System

Its disadvantages are

(a) Steady-state frequency to critical bus always below 60 Hz

(b) Transient frequency dip to approximately 57.5 to 58 Hz

(c) Only one chance for diesel start before frequency dips below 58 Hz

(d) Requires vibration mounts or sound insulation pads to prevent transmission of rotating equipment noise to other parts of building

(e) The diesel engine supplied with this system is limited to supplying power only for the critical bus and cannot be used to supply other standby power

(5) System 5 is shown in Fig 27. With this system, utility power is rectified to drive the direct-current motor and keep the batteries charged. The motor drives an alternating-current alternator with its own excitation and voltage regulator. The alternator supplies the critical bus. Upon loss of utility power, the stored energy in the battery will keep the motor generator set running with no disturbance in frequency or voltage of the critical bus supply. The amount of stored energy will depend on battery size. A 5 min capacity is usually ample for this type of system. In order to maintain continued supply of the critical bus during prolonged interruptions,

it is required to have a standby alternator set of sufficient capacity to recharge the battery and supply the full critical load. The standby alternator set must be complete with its own control and automatic transfer equipment. The size of the standby alternator will be determined by the requirements of the critical bus and battery and other essential loads.

The cost of this equipment, including a standby diesel alternator, will be approximately \$800 to \$900 per kilowatt, with 200 kW being the practical limit.

The advantages of system 5 are

(a) Moderate first cost

(b) Steady-state frequency is 60 Hz ± 0.25 Hz

(c) Can be used with standby diesel alternator set for other than critical bus function

Its disadvantages are

(a) Higher operating cost

(b) Maintenance of direct-current motor commutator

(c) Maintenance and ventilation of storage batteries

(d) Heavy weight of storage batteries

(e) Heavy weight of the motor generator set

(6) System 6 (Fig 28) is a steam turbine powered no-break configuration

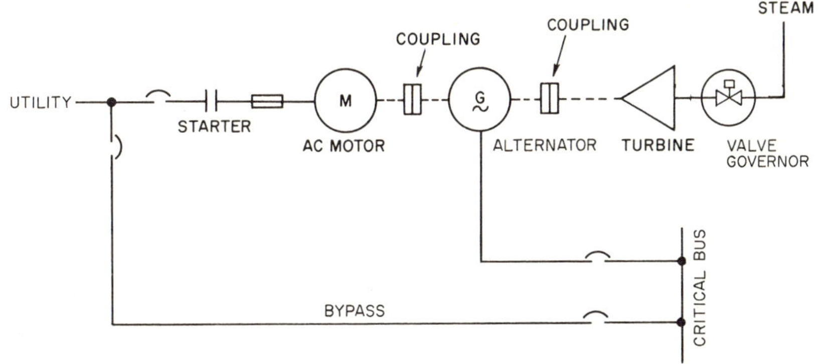

Fig 28
Steam-Turbine-Driven Emergency Power System

and consists of a squirrel-cage induction motor fed from the utility supply. It is used to drive a 60 Hz alternator which is complete with its own voltage control and excitation equipment. Coupled to the motor generator set is a simple steam turbine complete with governor to actuate the throttle control valve and a diaphragm-operated main steam valve. With failure of utility power the diaphragm admits steam to the throttle valve and thus to the steam turbine which becomes the prime mover. The valve system is supplied with a bypass to allow bleeding steam through the turbine at all times to keep it hot and avoid condensation.

With this system the rotor of the turbine is rotating at all times and the output speed of the turbine shaft matches that of the generator set which is normally driving it. Since the rotating mass does not have to be accelerated when utility power fails, the frequency deviation during changeover can be quite small.

The cost of this type of set is approximately $400 per kilowatt including all controls.

The advantages of system 6 are

(a) Low first cost

(b) Moderate maintenance costs

(c) No heavy battery or battery ventilation equipment required

Its disadvantages are

(a) Steady-state frequency to critical bus is always below 60 Hz and dips to approximately 57 Hz during changeovers

(b) Requires guaranteed supply of steam at all times to provide uninterruptible power when utility supply fails

4.7 Battery Systems

4.7.1 *Introduction.* A battery is the most dependable source available for emergency power, and when applied with other devices, can also be one of the most versatile. The application of batteries varies from the hand-held flashlight to the most sophisticated supply for computers.

4.7.2 *Codes, Rules, and Regulations Concerning Emergency Lighting.* Many states and municipalities have adopted codes requiring the use of emergency lighting. The National Electrical Code, NFPA No 70 (1975), Article 700, sets forth the standards of practice for emergency lighting equipment. The Life Safety Code, NFPA No 101 (1973), also provides details concerning emergency lighting.

**Fig 29
Typical Battery Unit**

Installation of emergency lighting aids in evacuating people from a building safely without injury or panic and may also keep a building or plant in limited operation. Exits, hallways, stairways, machinery, first-aid rooms, telephone exchanges, and operating rooms are just a few of the areas which may require or benefit from emergency lighting.

4.7.3 *Unit Equipment Emergency Lighting.* For evacuation use and use in vaults and switchgear rooms small inexpensive battery units are available. A typical unit is shown in Fig 29. These units contain a small charger to maintain a proper charge on the battery. In the event of an external power failure the lamps are automatically switched on and supplied from the battery. When external power is restored, the lamps are turned off and the batteries recharged.

Prices range from about $80 to $250 a unit, depending upon the battery type, number of lights, and length of maintained power during an interruption of regular power.

Similar units are available for use in fluorescent fixtures so that flush ceiling mounting decor may be maintained.

4.7.4 *Central Battery Lighting Systems.* Building-wide systems may also be used to connect many lamps to a central battery-charger console. Each of these approaches has basic advantages which should be considered.

For limited applications in relatively small areas 6 and 12 V emergency lighting units may be preferable. They can control as many as eight lamps from a single unit and can illuminate several rooms or areas simultaneously. Some require no battery maintenance during their service life, permitting their installation in areas where maintenance might become an awkward chore.

There may be need for more lamps and longer wire runs. The decision may be to install many 6 or 12 V units or a single 32 V or 115 V system with a power source (battery, charger, console) centrally located. The trend today, particularly in new construction where 10 to 100 lamps are involved, is to utilize the central system. The advantages of such a system are the following:

(1) Centralized power source, eliminating the need for single units located throughout the building, use of less space (only lamps are in the areas to be protected), and the power source is located in the power room, basement, etc. This facilitates maintenance and testing of the system.

(2) The availability of alarm and protection circuits which increase the flexibility of the system.

(3) The advantage of distributing at lower current, higher voltage, resulting in decreased losses.

(4) A central location reduces maintenance costs and provides a means for regular testing of the system for reliability.

When larger systems are justified, the 115 V system is recommended. Here, more lamps and longer wire runs are economical.

4.7.5 *Factors to Consider when Selecting Emergency Lighting System.* Immediate questions are, what area is to be protected, and what light level is necessary. The layout of the building, department, and usage of room will answer the question. Areas which should be lighted are places near moving machinery, passageways, exits, stairwells, etc. The light level required is dictated by several factors, including applicable codes, usage, available funds, and personal preference. The light level is determined by the number of lamps, their wattage, the lumen output of the lamps per watt, the reflector efficiency, the reflectance of the building surfaces, and the area covered.

As the number of lamps, length of wire runs, and size of lamps increase, the voltage of the system should be increased to obtain a minimum cost and a more efficient system. The 32 V system should be applied on intermediate-size applications; larger applications require the 115 V system.

In summary, there are three steps in sizing an emergency lighting system: (1) determine number, location, and type of lamps needed; (2) determine total lamp wattage of all selected lamps; and (3) determine protection time required.

Larger systems require a distribution panel. Its purpose is to distribute the available direct-current power to a number of individually fused circuits with each fuse preferably monitored by a visual and audible alarm circuit. If one circuit suffers a fault, the fuse protection removes the faulted circuit from the direct-current bus, allowing the battery to power the remaining circuits. The distribution panel is usually located in close proximity to the battery supply.

The normal alternating-current voltage supply to the battery charger and the battery voltage should be continually monitored. An alarm should be sounded whenever voltage fails and if the battery voltage should become high or low, indicating a malfunction of the battery charger. The battery system alarm may be in any convenient location.

4.7.6 *Multiple Sources Used for Normal Lighting.* The systems described up to now are used whenever a single voltage source is used for normal lighting. The load contactor senses the failure of the alternating-current voltage it monitors, and all emergency lamps will come on.

There are systems that will monitor each line-to-neutral voltage of a three-phase alternating-current system. Thus, upon failure of any one or all of the three voltages, the load contactor will close, turning on all emergency lights.

If it is not desirable to turn on all the emergency lights, but only the room or department where the normal lighting has failed, a multiload contactor panel may be used to monitor the alternating-current voltage that supplies the lighting in the area or zone protected by that load contactor. If that voltage should fail, then that contactor only will close, and its lights will come on. If more than one alternating-current voltage fails, then more than one string of emergency lights will come on.

When the normal lighting is supplied by mercury vapor or quartz iodine, full light is not available during the ignition period. This period could be as long as 5 to 10 min, depending on temperature and line voltage. The normal characteristic of the emergency lighting system is to supply lighting only when the normal power is not available, as soon as it

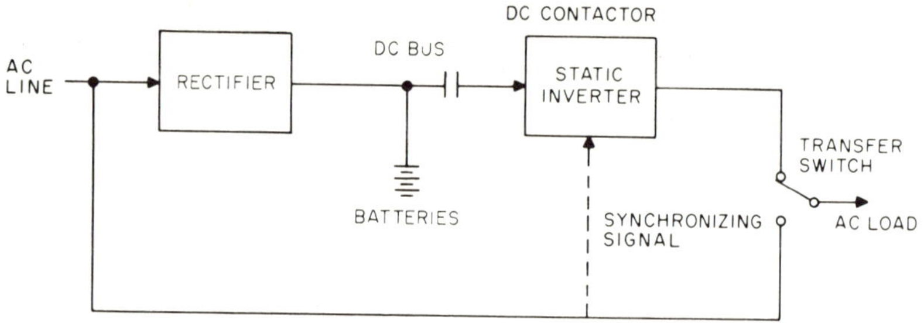

Fig 30
Short-Interruption Standby System

returns, the emergency lamps will go out. Thus there will be a period of time that no lighting will be available. To overcome this defect, a time delay on the load contactor should be considered.

The battery will try to continue to supply power as long as the normal power is not available. In actual practice, however, the battery is limited to the amount of power it can deliver. A power failure of long duration will result in an overdischarged battery and may cause permanent damage. To prevent this, a drop-out relay based on time, battery voltage level, or other system parameters is recommended.

4.7.7 Alternating-Current Power Supply with Mechanical Relay Transfer. Fig 30 shows a relay transfer system for supplying an emergency power source to a load upon the loss of the prime source. Power to the load will be interrupted, as shown in Fig 31, from 60 to 190 ms, depending upon the type and size of the transfer switch. More costly but faster transfer systems use static relays.

In the system of Fig 30 the loss of voltage on the alternating-current line will cause the direct-current contactor to close and supply power to the inverter. At the same time the transfer switch operates to transfer the load to the inverter supply. This system is adequate

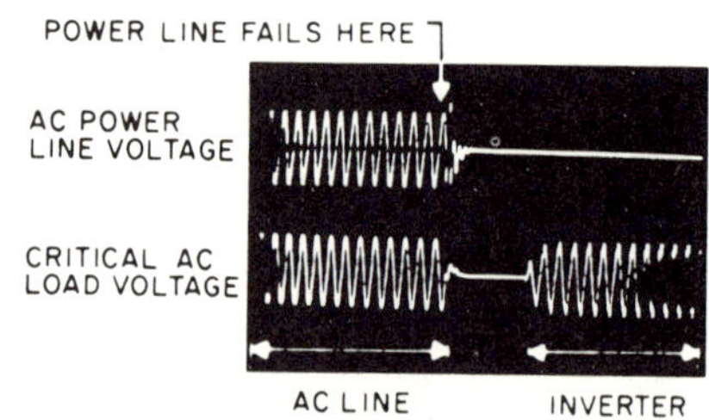

Fig 31
Oscillogram of Output Voltage of System of Fig 30 During Transfer

for lighting, signal circuits, radio systems, and other loads which can tolerate the short power failure.

4.7.8 Nonredundant Uninterruptible Power Supply System. One of the most widely accepted systems for loads which require alternating-current uninterruptible power are those consisting of a rectifier, battery, and inverter. These are available in sizes ranging from 250 VA to several hundred kilovolt-amperes. A typical one-line diagram of such a system is shown in Fig 32.

During normal operation, the prime power and rectifier supply power to the inverter and also charge the battery which is "floated" on the direct-current bus and kept fully charged. The inverter converts battery power from direct to alternating current for use by the critical loads. The inverter alone governs the characteristics of the alterna-

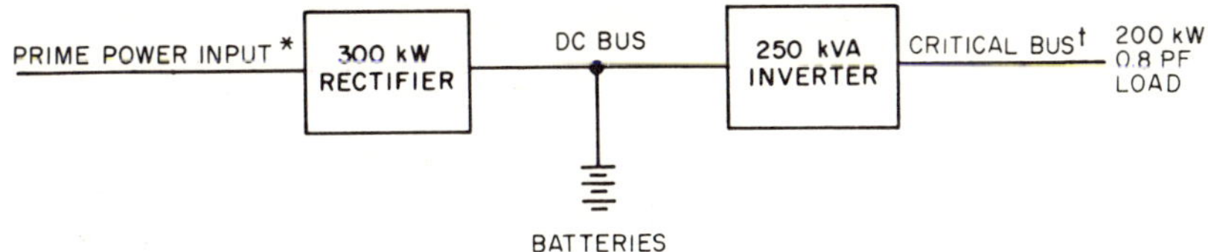

*May be subject to items (1)–(5):
†Is essentially free from items (1)–(5):

(1) Voltage dips
(2) Frequency variations
(3) Momentary interruptions
(4) Transient disturbances
(5) Power interruptions

**Fig 32
Nonredundant Uninterruptible Power Supply System**

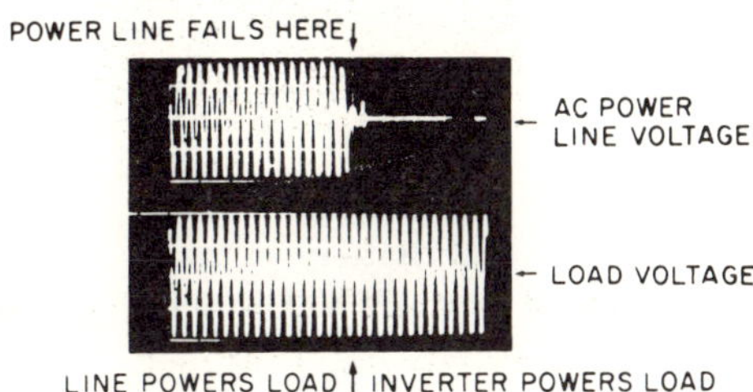

**Fig 33
Oscillogram of System of Fig 32 with
Power Line Failure**

ting-current output, and any voltage or frequency fluctuations or transients present on the utility power system are completely isolated from the critical load.

In the event of a momentary or prolonged loss of power, the battery (which is floated on the direct-current bus) will supply sufficient power to the inverter to maintain its output for a specified time until the battery has discharged to a predetermined minimum voltage.

Fig 33 shows an oscillogram of the continuous-load voltage supplied by the system of Fig 32, even when the prime source of power fails.

Upon restoration of the prime power, the rectifier section will again resume feeding power to the inverter and will simultaneously recharge the battery.

Static systems, as described, provide

(1) Precise uninterruptible power
(2) Low maintenance, no moving parts
(3) Easy installation, no special foundations needed
(4) High efficiency, static conversion devices
(5) Good performance, frequency unaffected by load changes, excellent voltage regulations, fast transient response

System availability should be as high as economically justifiable and may be calculated by using the following formula with all figures in the same units, usually hours:

$$A = \frac{MTBF}{MTBF + MTTR}$$

where

A = System availability

MTBF = Mean time between failures

MTTR = Mean time to repair

A typical specification for an uninterruptible power supply system is given in Table 12.

The nonredundant system shown in Fig 32 has the advantage of simplicity

Table 12
Typical 300 kVA Uninterruptible Power System Specifications

Input Power		*Battery System*	
Voltage	208Y/120 V ac	Battery type	Lead-calcium, nickel-cadmium, or lead-antimony
Power factor	0.85		
Frequency	60 Hz ±5 Hz		
Phases	3, 4 wire	Float voltage	241 V dc
Current limit	840 A rms	High rate voltage	250 V dc
Starting surge	0 to 100% load in 15 s	Finish voltage	1.75 V/cell (lead)
Harmonic insertion	30%		1.0 V/cell (nickel cadmium)
		Recharge time	10 × discharge time
		Capacity	Sized to requirement
Output Power			
Rating	300 kVA steady state	*General Characteristics*	
	450 kVA overload 15 s	Controls	Automatic line synchonization
Power factor	0.8		
Voltage	208Y/120 V ac		Circuit breakers: input, output, bypass, battery, and control power
Phases	3, 4 wire		
Frequency	60 Hz ±0.001%		
Voltage regulation	±1%		
Voltage adjustment	±5%	Instruments	Input voltmeter and ammeter
Total harmonic distortion	5%		Output voltmeter, ammeter, and frequency meter
Single harmonic content	3%		
Phase displacement	120° ± 1%		Battery voltmeter and ammeter
Load unbalance	30% of phase rating		
Fault current limit	1250 A rms per line	Alarms	Blower failure, open fuse, low battery
Transient voltage regulation			
1) Any rated load disturbance	±5% with 25 ms recovery	Efficiency	85%
		Environment	50°F to 110°F
2) Any input power failure or return	±5%		95% relative humidity
		Cooling	Blower cooled
3) Uninterrupted transfer to or from bypass power	±5%	System compatibility	Can be paralleled with like power supplies
		Weight	24 400 lb

and low cost. Large systems can have a predicted reliability of 20 000 h of mean time between failures using handbook reliability data; but field experience indicates that reliabilities in the order of 40 000 h of mean time between failures can be achieved. The nonredundant system has the disadvantage of disturbing the critical bus in the event of an inverter failure. This disadvantage can be overcome by the use of a redundant system as shown in Fig 34.

4.7.9 Redundant Uninterruptible Power Supply System. Fig 35 is an oscillogram of the two-inverter transfer system voltage during malfunction of one inverter similar to the arrangement shown in Fig 34. In the redundant system, each half of the system has a rating equal to the critical load requirements. The basic power elements (rectifier, inverter, and interrupter) are duplicated, but it is usually not necessary to duplicate the battery since its inherent reliability is extremely high. Certain control elements such as the frequency oscillator may also be duplicated.

The static interrupters isolate the faulty inverter from the critical bus and

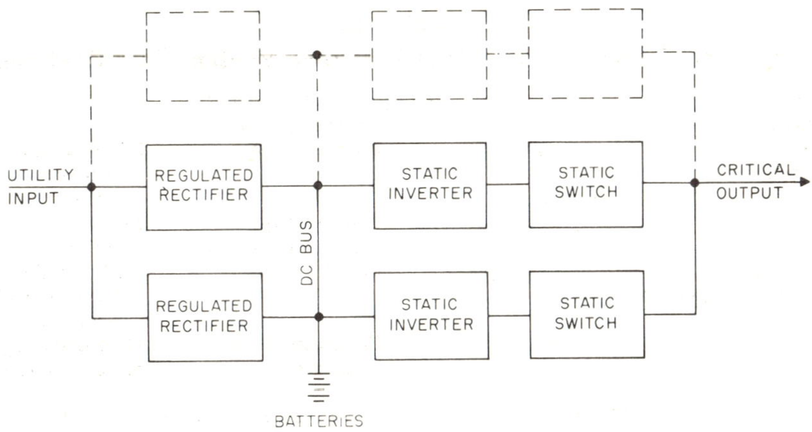

Fig 34
Redundant Uninterruptible Power Supply System

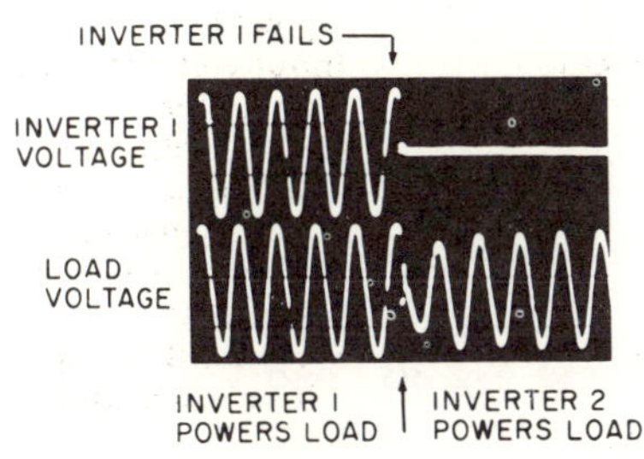

Fig 35
**Oscillogram of Output Voltage of System
of Fig 34 upon Inverter Failure**

prevent the initial failure from starting a "chain reaction" which might cause the remaining inverter to fail.

Large systems which require multiple rectifier/inverters to handle the load requirements would require one additional rectifier/inverter to provide redundancy. Redundant systems should consist of the fewest parallel paths required to supply the critical load requirements plus one additional path for redundancy. A larger number of smaller rated paths does not necessarily provide increased reliability since they add unnecessary additional components which are themselves subject to failure.

The cost of a redundant system is approximately $(N+1)/N$ greater than for a nonredundant system, where N equals the number of paths required for a nonredundant system. However, such a system is two to four times more reliable than a nonredundant system.

4.7.10 *Nonredundant Uninterruptible Power Supply System with Static Bypass Switch.* An alternate method of increasing the overall system reliability is a static bypass around the faulted inverter as shown in Fig 36. When an inverter fault is sensed, the critical load can be transferred to the bypass circuit in less than 5 ms. Fig 37 is an oscillogram of the load voltage as supplied during a transfer of the power source.

The static bypass adds about 20 percent to the cost of a nonredundant system, but is eight to ten times more reliable.

4.7.11 *Parallel Redundant Uninterruptible Power Supply System.* Fig 38 shows a parallel supplied parallel redundant uninterruptible power supply system. Reliability is a paramount consideration in this 1000 kVA system. Solid-state sensors and static switches, not shown, are installed to clear a malfunctioning inverter without effect on critical computer load.

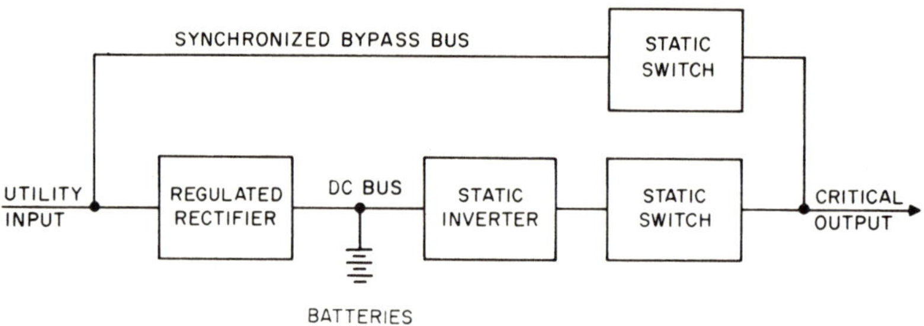

**Fig 36
Uninterruptible Power Supply System with Static Bypass**

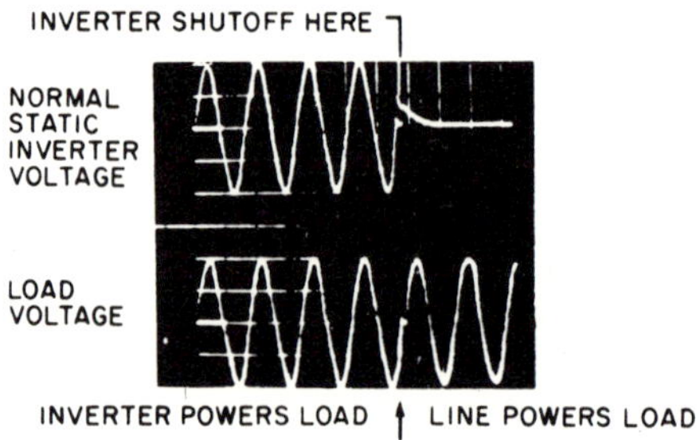

**Fig 37
Oscillogram of Static Switch of the System of Fig 36 Load Voltage**

Fig 39 shows an oscillogram of the voltage supplied to a load where the prime source of power is short circuited and a static transfer is made to a battery-powered inverter.

4.7.12 *Parallel Nonredundant Uninterruptible Power Supply System with Static Bypass Switch.* Fig 40 shows a parallel supplied nonredundant uninterruptible power supply system. The installation consists of two discrete systems serving two computers. A synchronized bypass and static switch protect each load in the event of inverter fault. Should voltage be lost to a computer load, the static transfer switch will operate to reestablish voltage in less than one quarter of a cycle, fast enough to be considered continuous power for most loads.

4.7.13 *Combination Static Inverter and Rotating Uninterruptible Power*

Supply System. A combination static, battery, and rotating uninterruptible power supply system is shown in Fig 41. In normal operation the rectifier is supplied from the prime power source, and the battery is floating on the line. The inverter output frequency is slaved to the prime source and follows it exactly. The solid-state rectifier is supplying direct-current power to the inverter and is also maintaining the batteries at the proper float charge. The generator supplying power to the load can be a standard commercial unit. No flywheels are used in this system.

During a fault in the incoming or normal source of power, the high-capacity battery supplies direct-current power to the inverter. The inverter frequency standard controls the frequency of the synchronous motor within $\pm$ 0.1 percent of rated standard while the battery

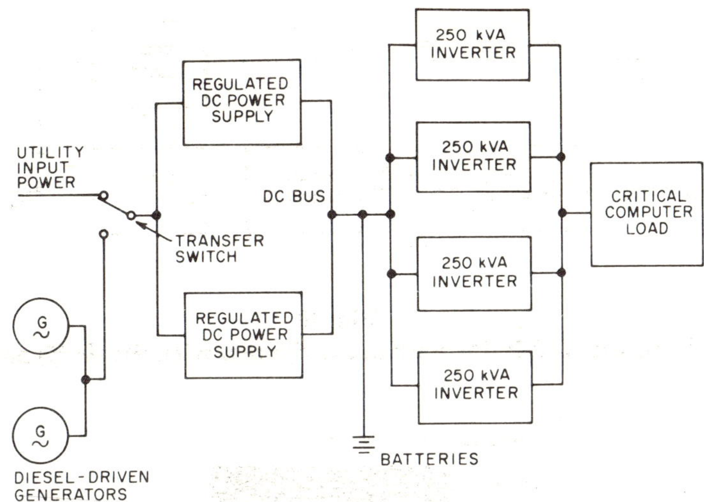

Fig 38
Parallel-Supplied, Parallel-Redundant Uninterruptible
Power Supply System

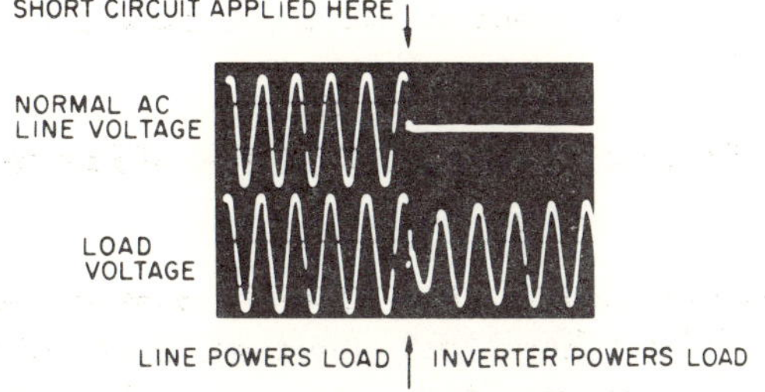

Fig 39
Oscillogram of Load Voltage Supply of System of Fig 38
upon Prime Source Short Circuit

is maintaining the system. Fig 42 illustrates the stability of the load frequency and voltage under widely varying input conditions. This system shows good voltage and frequency stability under starting loads, even with no base load.

4.7.14 *Combination Static Inverter and Engine Generator Uninterruptible Power Supply System.* A typical arrangement of a prime power source, an engine generator source, and a battery source of electric power is shown in Fig 43. During an interruption of normal utility power, the static uninterruptible power supply provides continuous power to critical selected loads. A time-delay relay prevents the engine generator from starting immediately. After a preset period, the engine generator automatically starts and when the voltage stabilizes, the automatic transfer switch connects the uninterruptible power supply load. The arrangement may be considered because of the possible simultaneous failure of both the electric and gas utilities.

The load is normally served by the inverter initially, but if uninterruptible

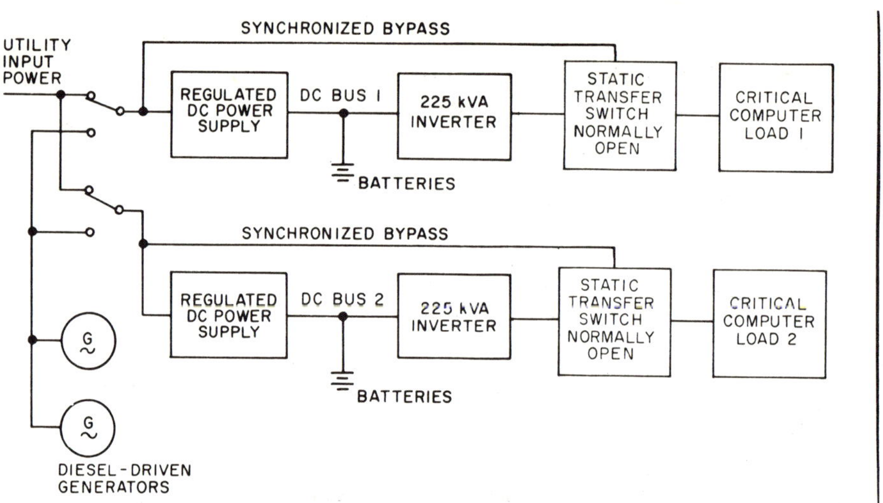

Fig 40
Parallel-Supplied Nonredundant Uninterruptible
Power Supply System

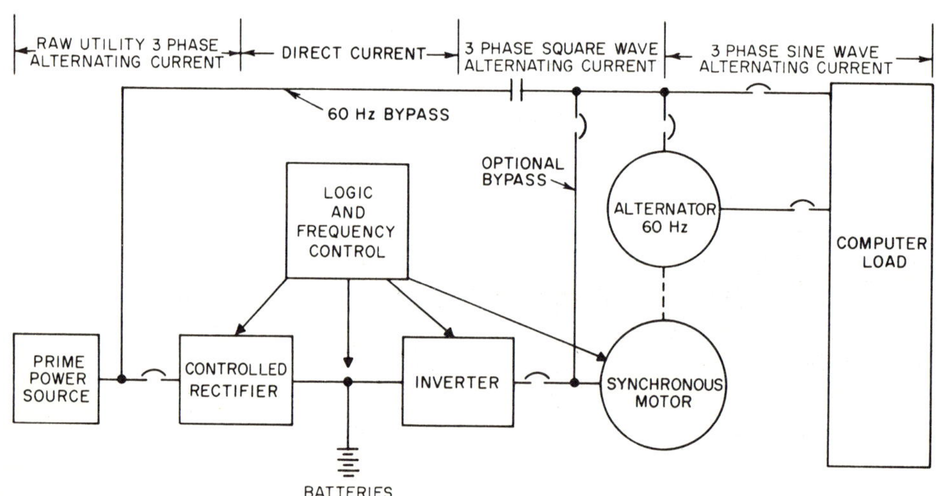

Fig 41
Combination Static, Battery, and Rotating Uninterruptible
Power Supply System

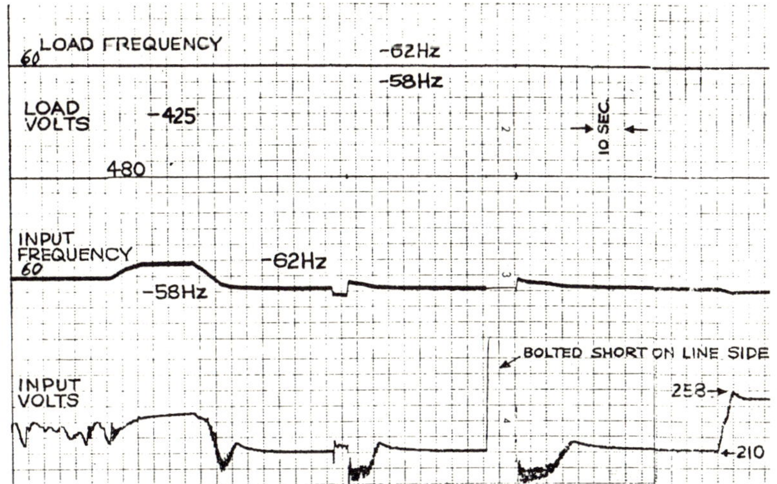

Fig 42
Load Frequency and Voltage Stability under
Varying Input Conditions

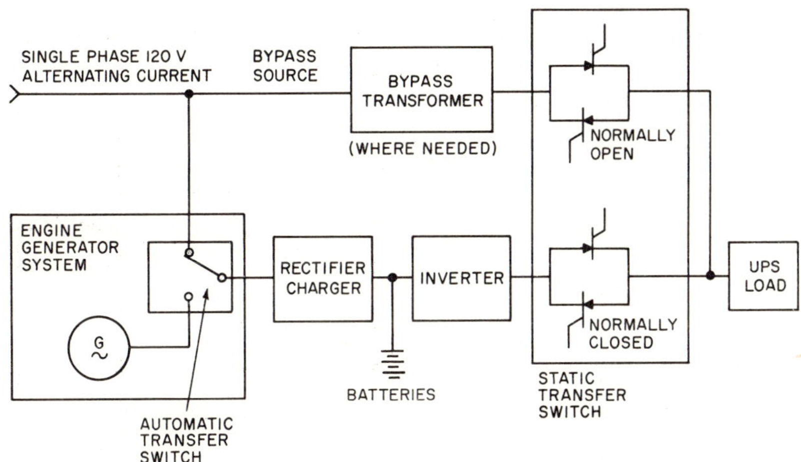

Fig 43
Typical Block Diagram of Combination Uninterruptible
Power Supply and Engine-Generator System

power supply equipment failure should occur, the load is transferred readily to the bypass source. The power ratings in kilowatts and the characteristics of both the engine generator and the uninterruptible power supply systems should be compatible.

4.7.15 *Battery Selection.* Inverters require a minimum direct-current voltage for commutation. A commutation failure results in a short circuit which must be cleared by the thyristor fuses. The minimum required direct-current voltage is usually 75 to 80 percent of nominal. A typical inverter nameplate reads 105 to 140 V direct-current input, 120 V alternating-current output. This must be considered when selecting battery sets for uninterruptible power supply systems. It is particularly important for applications where the battery must supply energy for short periods only, because of the relatively large current flowing through the internal resistance of the battery. Uninterruptible power supply systems should have a low-direct-current voltage alarm. Low-direct-current voltage trip of the inverter should be considered to prevent damage and useless discharging of the batteries.

Economy and reliability will be realized if the volt-ampere switching capabilities of thyristors are optimized with appropriate derating for reliability. The primary limitation on thyristors is current, that is, heating. Selection of proper battery voltage can be crucial. Optimum battery voltages based on design tradeoffs are given in Table 13.

The battery is sized to support the critical load until either (1) the critical load can be shut down in an orderly manner or (2) the utility power returns or an alternate standby prime source can be started and connected. Typical battery support times might be 5, 15, or 30 min. Rather than purchasing larger

**Table 13
Optimum Battery Voltages**

Output (kVA)	Optimum Battery Voltage (V dc)
Three phase	
10–30	105–140
37.5–100	210–280
150*–1000	320–430
Single phase	
2.5–10	105–140
15–30*	210–280
37.5–60*	210–280

*A three-phase inverter is more economical and has other advantages at high power.

battery capacity, an engine or turbine generator standby power source should be considered.

Nearly all storage batteries used in emergency power systems today are of either the lead-acid or the nickel-cadmium type. Each type has its advantages and disadvantages, as summarized in Table 14.

The battery system should be sized from manufacturers' data for a particular application in a known operating temperature range. Most ampere-hour ratings are for a temperature of 77°F, and a reduction of ampere-hour capacity is usually necessary for operation at lower temperatures. Some manufacturers derate their lead-acid batteries by as much as 60 percent from the 77°F ratings for operation at 0°F.

The batteries for an uninterruptible power supply system imposing a continuous load may be sized by multiplying the load current in amperes by the time in hours to determine the number of ampere-hours required. However, ampere-hour capacities decrease as the rate of discharge increases. Thus if the load is varying, the battery capacity required will be the summation of the various loads,

$$A \cdot h = A_1 t_1 + A_2 t_2 + \ldots + A_n t_n$$

Table 14
Battery Characteristics

Battery Type	Composition	Typical Characteristics
Lead-antimony	Lead and lead antimony plates; sulfuric acid electrolyte	Life approximately 14–18 years; rated at 77°F, with operating range from 10°F to 110°F. Capacity decreases with decrease in temperature; requires periodic equalizing charge. Lowest in initial cost.
Lead-calcium	Lead and lead calcium plates; sulfuric acid electrolyte	Life approximately 20–30 years; same temperature ratings as lead-antimony; does not require periodic equalizing charge if floated between 2.20 and 2.25 V per cell. Cost approximately 15 percent more than lead-antimony.
Nickel-cadmium	Cadmium and nickel oxide plates; potassium hydroxide electrolyte	Life in excess of 25 years; better low temperature capacity; very little gassing while charging; not damaged by complete discharge. Highest in cost; as much as two to three times the cost of lead-antimony for the same ampere-hour capacity; requires periodic equalizing charge.

where

$A \cdot h$ = Ampere-hours

A = load, in amperes

t = Time, in hours

One additional factor to consider is whether the battery will deliver the capacities in the order required. Usually, a larger capacity battery will be required if there is a large discharge rate at the end of the cycle. To verify that the battery selected is adequate, it should be checked by starting at the beginning of the discharge cycle and subtracting the energy removed At by each load in order to determine if adequate capacity still remains for the final load interval.

Experience has shown that the lead-calcium battery requires several days to weeks to return to full equal charge on all cells following a discharge to the minimum inverter input voltage. Other battery types may be required where frequent prolonged power outages occur, since there may not be time to fully charge between outages without raising the charging voltage beyond the rating of the inverter. Long life with a minimum of service and infrequent demands for power indicates advantages for remote locations.

A typical battery installation for emergency power is shown in Fig 44. Such installations may supply direct-current power directly or may be used in conjunction with direct-current motors driving alternators or connected to an inverter for an alternating-current supply.

4.7.16 *Battery Charger*. The charging rectifier, or battery charger, is a very important part of the emergency power system, and consideration should be given to redundant chargers on critical systems. A general formula for sizing the battery charger for an inverter system would be as follows:

battery charger (amperes)

$$= \frac{\text{inverter output (VA)} \times 100}{\text{voltage input} \times \text{conversion efficiency}}$$

$$+ \frac{1.15 \times \text{battery bank capacity (A·h)}}{\text{desired recharge time (hours)}}$$

Fig 44
Typical Battery, Rated 60 V, 1800 A . h, at 8 h Discharge Rate

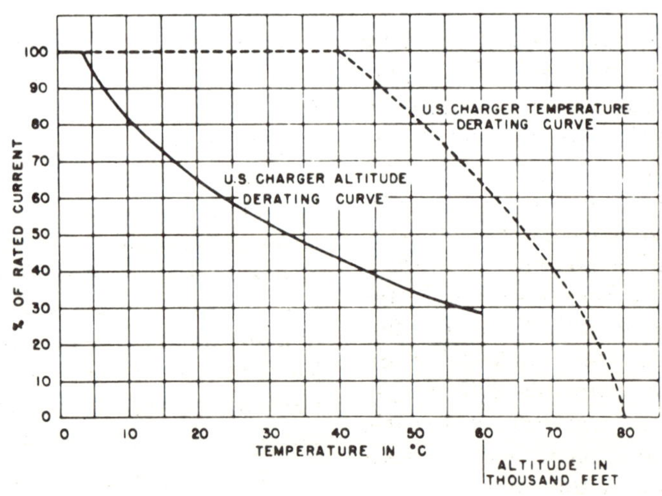

Fig 45
Derating Curves for Battery Chargers Due to Altitude and
Temperature

Table 15
1972 Prices for "Package" Systems
Without Batteries

Output (kVA)	Budgetary Price (dollars)	Cost per kVA (dollars
10	12 400	1240
150	69 000	460
300	120 000	400

Table 16
1971 Prices for Batteries Only

| Output (kW) | Support Time | | | |
	5 min	15 min	30 min	60 min
25	3600	4000	5000	6500
125	12 700	16 100	18 900	27 400
250	23 700	27 400	34 200	48 300
500	39 700	48 400	56 000	67 800
1000	59 800	71 400	89 600	135 600
2000	119 600	142 800	179 300	268 900

Table 17
Estimated Cost of "Package"
with Batteries

Output (kVA)	Battery Supply (min)	Budgetary Price (dollars)
125	15	100 000
125	30	104 000
125	60	113 000

The battery charger output must be derated for both altitude and temperature. These requirements have to be recognized since the user frequently establishes these conditions. A larger than normal rating may be required to make up the reduced capacity. A typical derating graph is shown in Fig 45.

4.7.17 *Cost of a Nonredundant Uninterruptible Power Supply System.* Budgetary 1972 prices for "package" nonredundant uninterruptible power systems with rectifier, inverter, and static switch, but without batteries or installation, are given in Table 15.

A typical midwestern U.S. 10 kVA system with ½ h battery supply, battery rack, and using union electrical contract labor, installed and operating complete in place, costs approximately $1800 to $2100 per kilovolt-ampere (1973).

4.7.18 *Cost of Batteries.* Typical 1971 estimating prices for batteries less installation costs are given in Table 16.

A complete static "package" with batteries, but without installation costs, is given in Table 17.

4.7.19 *Continuity of Operation Alarms.* In any areas involving battery and charger, supervision should be given serious consideration. A standby system which must be depended on, of necessity, should have audible and visual alarms to indicate such malfunctions as low battery voltage, loss of alternating-current power, high battery voltage, and low electrolyte level.

4.7.20 *Additional Information.* Additional information on uninterruptible power supply systems utilizing batteries, chargers, and inverters is given in [14]–[20].

4.7.21 *Special Precautions.* Extreme care must be exercised when specifying, purchasing, and applying an uninterruptible power supply system with a solid-state inverter serving multiple loads. Load changes due to switching a system off or on, or an automatic transfer of one of the loads may cause disturbances in voltage which are unacceptable by the other loads.

Special attention should be given to this problem which may be solved by one or a combination of the following:

(1) An uninterruptible power supply system may be purchased which maintains an acceptable voltage supply for any step load change. This may take the form of a higher capacity supply on which any variable load connected is a small percentage of the total capacity, or an extremely rapid voltage regulator capable of maintaining an acceptable voltage level.

(2) Loads should be purchased that have power fail detection hardware and a software routine which take the necessary protective action to provide the required on-line system reliability.

5. Case Histories

5.1 Introduction. Case histories of user needs for emergency or standby power systems are documented in this section. Preventative maintenance and full-load emergency operation simulation to test the equipment is required for reliable operation of an emergency or standby power system. Several examples follow.

5.2 Determining User Needs

(1) A survey of earthquake damage to buildings was conducted following recent tremors in San Fernando, California. 174 elevators failed because electric generators were made inoperative.

(2) A major manufacturing concern installed a voltage monitor on the input system to a computer room late in 1973. The input voltage was normally 208 V, alternating current, three phase, 60 Hz. In five weeks seven transients were recorded. Six of the transients were voltage drops of more than 10 percent, one was a voltage "spike" of more than 10 percent. Four of the transients exceeded 8 ms in length; three were immediately followed by computer malfunctions. Four of the transients occurred during a Sunday evening hour. The data collected showed that the voltage transients were troublesome to operations and justified a study to overcome the problem.

5.3 Lack of Maintenance

(1) An air base was immobilized for over 3 h after a storm caused a power interruption on the utility lines and the off-line standby diesel driven generator would not start. The utility company had to complete temporary repairs before the expensive standby equipment at the air base was operational.

(2) Lightning caused an interruption on a utility line which immobilized a draw bridge in a partly closed position. In addition, the radio control for the ships entering the channel went out of service. The standby gasoline engine driving a generator failed to start because the gasoline had evaporated.

(3) A "mechanical stored-energy system" was installed so that critical combustion controls for heating highly volatile solvents would ride through the momentary outage resulting from the utility switching from normal to emergency feeders. Although the equipment was designed and built by a reputable electrical equipment manufacturer, the first year's maintenance was so high and the interruptions caused by the system occurred so often that operating personnel installed jumpers to bypass the equipment. The utility supply was more reliable than the buffer equipment.

5.4 Misapplications of Emergency or Standby Power Systems

(1) Illustrative of the kind of problem which can result when computer systems are operated directly from engine generator set power is an airport where diesel engine generators were installed to back up the flight control tower functions. When the utility supply failed during the great Northeast power blackout, these diesel engine generators were successfully started and alternating-current power from the sets was applied to the computer bus within a few minutes. But as a result of the short-duration loss of input alternating-current power and the voltage and frequency transients encountered in the change-over to diesel engine generator set power, the computer system suffered extensive damage which took several days to repair on an around-the-clock emergency basis during which time the control tower and flight operations had to be operated on a substantially reduced basis.

(2) Engine-driven generators with a manual starter are sometimes used on farms. Upon loss of the utility electric power at the remote location the engine is started and the electricity is supplied to the farm system by a cord, plugged into an outlet. Opening of the main supply switch at the meter is required to prevent feedback to other customers and for the prevention of damage to the farmer's generating equipment when the utility power is restored. Such switching methods are hazardous and unsafe since they involve human error and forgetfulness. Automatic load switching should be provided, or as a minimum, a double-throw manual switch of sufficient capacity to transfer the total load from one source of power to the other should be installed.

5.5 Successful Operation of Emergency or Standby Power Systems

(1) During a massive blackout in which a large utility lost its entire system, bridges and tunnels to an adjacent state supplied by another utility continued to operate because the alternate source was from the other utility which continued to operate. Bridges and tunnels supplied by two sources from the utility which lost its system were without power.

(2) Two uninterruptible power systems, one located in Pasadena and one in Los Angeles, rode through the 1970 earthquake and worked perfectly when the utility company power was lost due to the quake. Both systems were installed to supply uninterruptible power to extensive computer systems.

6. References, Bibliography, and Working Groups

6.1 Standards References. The following standards publications and laws were used as references in preparing this document.

ANSI C84.1-1970, Voltage Ratings for Electric Power Systems and Equipment (60 Hz)[1]

EGSMA EGS1-1970, Standard Specifications for Standby Engine Driven Generator Sets[2]

EGSMA GDT2-1971, Glossary of Standard Industry Terminology and Definitions[2]

EGSMA IMFS1-1974, Standards and Recommendations for Installation and Maintenance of Farm Standby Electric Power[2]

EGSMA TDGS1-1972, Standard Specifications for Tractor Driven Generator Sets[2]

IEEE Std 100-1972, Dictionary of Electrical and Electronics Terms (ANSI C42.100-1972)[3]

IEEE Std 141-1969, Electric Power Distribution for Industrial Plants[3]

IEEE Std 142-1972, Grounding of Industrial and Commercial Power Systems (ANSI C114.1-1973)[3]

IEEE Std 241-1974, Electric Power Systems in Commercial Buildings[3]

IEEE Std 357-1973, Protective Relaying of Utility–Consumer Interconnections[3]

IEEE Std 387-1972, Criteria for Diesel-Generator Units Applied as Standby Power Supplies for Nuclear Power Generating Stations[3]

NECA Electrical Design Library Series 17, Electrical Design Guidelines (1971)[4]

NECA Electrical Design Library Series No 3/74, Emergency and Standby Power Generation (1974)[4]

NFPA No 70, National Electrical Code (1975), (ANSI C1-1975)[5]

NFPA No 76A, Essential Electrical Systems for Health Care Facilities (1973)[5]

NFPA No 101, Life Safety Code (1973)[5]

Public Law No 91-596, Williams-Steiger Occupational Safety and Health Act of 1970 (84 Stat. 1593, 1600; 29

---

[1]American National Standards Institute, 1430 Broadway, New York, NY 10018.

[2]Electrical Generating Systems Marketing Association, Tribune Tower, 435 N Michigan, Chicago, IL 60611.

[3]Institute of Electrical and Electronics Engineers, Inc, 345 East 47 Street, New York, NY 10017.

[4]National Electrical Contractors Association, 7315 Wisconsin Avenue NW, Washington, DC 20014.

[5]National Fire Protection Association, 470 Atlantic Avenue, Boston, MA 02210.

U.S.C. 656, 657), Chapter XVII of Title 29 of The Code of Federal Regulations, established on April 13, 1970 (36 F.R. 7006) as amended by adding thereto a new part 1910. Washington, DC: Department of Labor.[6]

Amendments to Williams-Steiger Occupational Safety and Health Act of 1970, as published in the Federal Register. Washington, DC: Office of the Federal Register, National Archives and Records Service, General Services Administration.[6]

Title 24, California Administrative Code, Part 3, Basic Electrical Regulations, Article E700, Emergency Systems. Sacramento, CA: Document Section.[7]

6.2 References

[1]   *Readers Digest Almanac.* Pleasantville, NY: The Readers Digest Association, 1972, p 790.

[2]   CASTENSCHIOLD, R. Criteria for Rating and Application of Automatic Transfer Switches. *IEEE Conference Record of 1970 Industrial and Commercial Power Systems and Electric Space Heating and Air-Conditioning Joint Technical Conference,* IEEE 70C8-IGA, pp 11-16.

[3]   KAUFMAN, J. E., Ed. *IES Lighting Handbook.* New York: Illuminating Engineering Society, 1972, sec 14, pp 14-10 to 14-12 (Emergency Lighting).[8]

[4]   *Power Engineering* (Power Circuit), Jan 1972, pp 12, 14.[9]

[5]   FINK, D. G., and CARROLL, J. M. *Standard Handbook for Electrical Engineers,* 10th ed. New York: McGraw-Hill, 1968, pp 20-8, 20-9.

[6]   GILBERT, M. M. What the Chemical Industry Can Do to Minimize Effects from Electrical Disturbances. Paper CP 60-1161, presented at the AIEE and ASME National Power Conference, Philadelphia, PA, Sept 21-23, 1960.

[7]   IEEE Committee Report. *Protection Fundamentals for Low-Voltage Electrical Distribution Systems in Commercial Buildings.* IEEE JH 2112-1, 1974, pp 19, 20.

[8]   KATZ, E. G. Evaluation of Hospital Essential Electrical Systems. *Fire Journal,*[5] vol 62, Nov 1968.

[9]   MIROWSKY, B. J. Effects of Short Duration Power Interruptions on the Computerized Checkout System for NASA. *Proceedings of the American Power Conference,* vol 28, 1966, pp 1049-1067.

[10] ALLINGHAM, R. E. Power for Computers. Presented at the Annual Convention Meeting of the Electrical Generating Systems Marketing Association, Harbor Springs, MI, Sept 22-27, 1970.

[11] IEEE Committee Report. Reliability of Electrical Equipment, Pt 1. *IEEE Transactions on Industry Applications,* vol IA-10, Mar/Apr 1974, pp 213-235.

[12] SAWER, J. W. Gas Turbine Emergency/Standby Power Plants. *Gas Turbine International,*[10] Jan/Feb 1972.

[13] HEISING, C. R., and JOHNSTON, J. F., JR. Reliability Considerations in Systems Applications of Uninterruptible Power Supplies. *IEEE Transactions on*

---

[6]Superintendent of Documents, US Government Printing Office, Washington, DC 20402.

[7]Document Section, PO Box 20191, Sacramento, CA 95820.

[8]345 East 47 Street, New York, NY 10017.

[9]1301 South Grove Avenue, Barrington, IL 60010.

[10]80 Lincoln Avenue, Stamford, CT 06904.

Industry Applications, vol IA-8, Mar/Apr 1972, pp 104-107.

[14] KUSKO, A., and GILMORE, F. E. Concept of a Modular Static Uninterruptible Power System. *Conference Record of the 1967 IEEE Industry and General Applications Group Annual Meeting,* IEEE 34C62, pp 147-153.

[15] LAWSON, L. J. A True No-Break, Off-Line Uninterrupted Power Supply. *Conference Record of the 1967 IEEE Industry and General Applications Group Annual Meeting,* IEEE 34C62, pp 154-158.

[16] GRIFFITH, D. C., and YUEN, M. H. Static No-Break Power for Critical Loads in a Modern Oil Refinery. *Conference Record of the 1967 IEEE Industry and General Applications Group Annual Meeting,* IEEE 34C62, pp 643-652.

[17] KUSKO, A., and GILMORE, F. E. Application of Static Uninterruptible Power Systems to Computer Loads. *Conference Record of the 1969 IEEE Industry and General Applications Group Annual Meeting,* IEEE 69-C5 IGA, pp 635-639.

[18] RELATION, A. E. UPS Systems for Critical Power Supplies. *Conference Record of the 1971 IEEE Industry and General Applications Group Annual Meeting,* IEEE 71C1-IGA, pp 877-884.

[19] WALKER, L. H. Inverter for UPS with Subcycle Fault Clearing Capabilities. *Conference Record of the 1971 IEEE Industry and General Applications Group Annual Meeting,* IEEE 71C1-IGA, pp 361-370.

[20] WOLPERT, T. Uninterruptible Power Supply for Critical AC Loads — A New Approach. *Con-ference Record of the 1973 IEEE Industry Applications Society Annual Meeting,* IEEE 73CHO763-3IA, pp 595-602.

6.3 Bibliography

[21] BEEMAN, D. L., Ed. *Industrial Power Systems Handbook.* New York: McGraw-Hill, 1955.

[22] BURCH, B. F., JR. Protection of Computers against Transients, Interruptions, and Outages. Presented at the 1967 IEEE Industry and General Applications Group Annual Meeting, Pittsburgh, PA, Oct 4, 1967.

[23] FERENCY, N. Is a UPS Really Necessary? *Electronic Products Magazine,* [11] Apr 17, 1972, pp 126-127.

[24] FISCHER, E. I. Emergency Power Facilities for a Research Laboratory. *Conference Record of the 1966 IEEE Industry and General Applications Group Annual Meeting,* IEEE 34-C36, pp 305-324.

[25] GROSS, S. Rapid Charging of Lead Acid Batteries. *Conference Record of the 1973 IEEE Industry Applications Society Annual Meeting,* IEEE 73CHO763-3IA, pp 905-912.

[26] HAUCK, T. A. Motor Reclosing and Bus Transfer. *IEEE Transactions on Industry and General Applications,* vol IGA-6, May/Jun 1970, pp 266-271.

[27] HEISING, C. R., and DUNKI-JACOBS, J. R. Application of Reliability Concepts to Industrial Power Systems. *Conference Record of the 1972 IEEE Industry Applications Society Annual Meeting,* IEEE 72CHO685-81A, pp 289-295.

----
[11] 645 Stewart Avenue, Garden City, NY 11530.

[28] HELMICK, C. G. Designing for System Reliability in Large Uninterruptible Power Supplies. *Conference Record of the 1971 IEEE Industry and General Applications Group Annual Meeting,* IEEE 71C1-IGA, pp 371-384.

[29] HELMICK, C. G. Uninterruptible Power Supply Systems — What, Why, Where, and When? Presented at the 34th American Power Conference, Chicago, IL, Apr 18-20, 1972.

[30] KATZAROFF, P. A Base Guide to Uninterruptible Power Systems. *IEEE Conference Record of the 1974 26th Annual Conference of Electrical Engineering Problems in the Rubber and Plastics Industries,* IEEE 74CHO831-8IA, pp 1-6.

[31] KENNY, R. W., McGOVERN, M. J., and TORPEY, P. J. Development of a Gas Turbine-Alternator System for Emergency Power Applications. *IEEE Transactions on Industry and General Applications,* vol IGA-1, Jan/Feb 1965, pp 3-8.

[32] KNIGHT, R. L., and YUEN, M. H. The Uninterruptible Power Evolution — Are Our Problems Solved? *Conference Record of the 1973 IEEE Industry Applications Society Annual Meeting,* IEEE 73CHO763-3IA, pp 481-485.

[33] LAWSON, L. J. New Uninterruptible Power System Alternatives Using High Capacity Kinetic Energy Wheels. *Conference Record of the 1973 IEEE Industry Applications Society Annual Meeting.* IEEE 73CHO763-3IA, pp 151-156.

[34] MILLER, N. A. Noninterruptible Power Supplies for Essential Control Systems in Power Plants. Presented at the IEEE IAS – IEC Group Meeting on Heating, Chicago Chapter, Jun 7, 1972.

[35] OLIVER, R. L. Plant Engineering at RCA. *Plant Engineering,*[9] Apr 1969.

[36] PALKO, E. Standby Generator Specification Chart. *Plant Engineering,*[9] Feb 18, 1971, pp 65-70.

[37] REICHENSTEIN, H. W., and CASTENSCHIOLD, R. Coordinating Overcurrent Protective Devices with Automatic Transfer Switches in Commercial and Institutional Power Systems. *IEEE Conference Record of the 1974 Industrial and Commercial Power Systems Technical Conference,* IEEE 74CHO855-7IA, pp 89-104.

[38] RELATION, A. E. UPS Systems for Critical Power Supplies. *Conference Record of the 1971 IEEE Industry and General Applications Group Annual Meeting,* IEEE 71C1-IGA, pp 877-884.

[39] RELATION, E. A., WINPISINGER, J. L., and MITCHELL, J. T. Uninterruptible Power System Using an Improved Magnetic Voltage Stabilizer. *Conference Record of the 1973 IEEE Industry Applications Society Annual Meeting,* IEEE 73CHO763-3IA, pp 17-23.

[40] RENFREW, R. M. Successful Uninterruptible Power Systems for Computers. *Conference Record of the 1968 IEEE Industry and General Applications Group Annual Meeting,* IEEE 68C27-IGA, pp 787-792.

[41] ROBERTS, A. M. Power Failure Ride-Through for an Inverter System Using Its Own Induction Motor Load as the Energy Source. *Conference Record of the*

1968 IEEE Industry and General Applications Group Annual Meeting, IEEE 68C27-IGA, pp 737-742.

[42] SCHWARM, E. G., and LITTLE, A. D. Computer Uninterruptible Power System with High Speed Static Bypass. Presented at the Summer Power Meeting and International Symposium of High Power Testing of the IEEE Power Engineering Society, Portland, OR, Jul 18-23, 1971.

[43] SUMMERS, G. E. Providing Reliable Power for Computer Systems. *Plant Engineering,* [9] Jan 7, 1971.

[44] SWENSON, E. C. How to Select and Install Standby Electric Plants. *Electrical Construction and Maintenance,* [12] Jan 1963.

[45] The Exciting World of Rechargeable Batteries. *Factory,* [13] Apr 1967, pp 84-87.

[46] System for Orderly Emergency Shutdown. *Modern Manufacturing,* [12] Dec 1969.

[47] Beating the Blackouts. *The Wall Street Journal,* [14] Jul 21, 1970.

[48] The On-Site Power Market. *Electrical Construction and Maintenance,* [12] Jan 1971, pp 57-71.

[49] Uninterruptible Power System Prevents Computer Downtime. *Rubber World,* [15] Nov 1970, pp 58-60.

[50] The Electric Way to Standby Power. *Plant Operating Management,* [16] Feb 1970, pp 62-65.

[51] Emergency and Standby Power Systems. *Electrical Consultant,* [17] Oct 1971.

[52] The Automatic Transfer Switch Heat of Emergency Power. A Reliability Study of a Power Supply System. The Battery World. *Electrical Consultant,* [17] vol 88, Nov 1972.

6.4 Manufacturers' Data

Rating Factors for Generating Plants. Tech Bull T-917. ONAN Company, 1400 73rd Avenue NE, Minneapolis, MN 55432.

TERVAY, J. C. Nickel Cadmium Pocket Plate Batteries for Standby Power Applications and Systems. Nife, Inc, 23 Dixon Avenue, Copiague, NY 11726.

Standby Gas Turbine Alternator Package. Publ SD1984. International Harvester Company, 2200 Pacific Highway, San Diego, CA 92112.

Synchronizer. Publ 200-Syn-68 (Gas Turbine). Electric Machinery Manufacturing Company, Minneapolis, MN 55413.

Emergency Lighting Handbook. Radiant Industries, Inc, 10900 Burbank Boulevard, North Hollywood, CA 91601.

6.5 National Directories for Products and Corporations

Conover-Mast Purchasing Directory. Conover-Mast, 95 East Putnam Avenue, Greenwich, CN 06830.

MacRae's Blue Book, 4 vols. MacRae's Blue Book Company, 100 Shore Drive, Hinsdale, IL 60521.

The National Buyers' Guide. Reuben H. Donnelley Corporation, 825 Third Avenue, New York, NY 10022.

---

[12] 1221 Avenue of the Americas, New York, NY 10020.
[13] PO Box 8785, Philadelphia, PA 19101.
[14] PO Box 300, Princeton, NJ 08540.
[15] 77 N Miller Road, Akron, OH 44313.
[16] 270 St Paul Street, Denver, CO 80206.

[17] 1760 Peachtree Road NW, Atlanta, GA 30309.

Sweet's Industrial Catalog System, 11 vols (vol ELC — electrical). McGraw-Hill Information Systems Company, 1221 Avenue of the Americas, New York, NY 10020.

Thomas Register of American Manufacturers, vols 1–6 Products and Services, vol 7 Company Names, Addresses, vol 8 Brand Name Index, vols 9–11 Catalog of Companies (alphabetical). Thomas Publishing Company, 1 Pennsylvania Plaza, New York, NY 10001.

6.6 Working Groups, Committees, and Societies Working on Similar or Complementary Projects

Canadian Standards Association (CSA), Standards Division, 178 Rexdale Boulevard, Rexdale, Ontario; Committee on Electrical Power Supply for Emergency Systems.

Computer and Business Equipment Manufacturers Association (CBEMA), 1828 L Street NW, Washington, DC 20036.

Electrical Generating Systems Marketing Association (EGSMA), Tribune Tower, 435 N. Michigan, Chicago, IL 60611; Technical and Standards Committee.

Institute of Electrical and Electronics Engineers (IEEE); Power Engineering Society, Transmission and Distribution Committee, General Systems Subcommittee, Working Group on Service to Critical Loads.

Instrument Society of America (ISA), 400 Stanwix Street, Pittsburgh, PA 15222; Committee on Emergency Power Supplies SP54.

National Electrical Contractors Association (NECA), 7315 Wisconsin Avenue NW, Washington, DC 20014

Acknowledgment

The Institute of Electrical and Electronics Engineers, Inc, wishes to express its appreciation to the following individuals and organizations for permission to include their copyrighted material in this document.

I. E. Trost, Inverter Systems, Westinghouse Electric Corporation, Buffalo, NY 14240

Frank O. Jappel, Manager Product Marketing, North American Turbine Corporation, Houston, TX 77040

H. L. Banbury, Associate Director, Electrical Generating Systems Marketing Association, Chicago, IL 60611

D. H. Roll, Vice President-Finance, Cummins Power, Inc, Commerce City, CO 80022

Ron A. Lang, Turbomachinery Sales, Solar Division of International Harvester Company, San Diego, CA 92138

Gary R. Squires, Manager of Advertising and Public Relations, Gould, Inc, Trenton, NJ 08607

R. R. Strong, I. W. Strong and Associates, Inc, Denver, CO 80204

Gas Turbine International, Stamford, CT 06904